COALS O

or

The Reason of My Wrighting

The autobiography of
ANTHONY ERRINGTON,
a Tyneside colliery waggon and waggonway wright,
from his birth in 1778 to around 1825

edited
by
P.E.H. HAIR

with illustrations, mainly by Thomas Bewick

published for the
DEPARTMENT OF HISTORY
UNIVERSITY OF LIVERPOOL

LIVERPOOL UNIVERSITY PRESS
1988

IN DEBT TO MY PARENTS
born Chirton, Tyne and Wear, and Coundon, County Durham

AND TO MY LANDLADY AND MY WORKMATES ON THE HAULAGE
Linton Colliery, Northumberland, 1944-1947

Liverpool Historical Studies, no. 3
General Editor: P.E.H. Hair

Published in Great Britain, 1988,
by the Liverpool University Press,
Senate House, P.O. Box 147, Liverpool L69 3BX

British Library Cataloguing in Publication Data are available

ISBN 085323 266 0

Printed in Great Britain by Manchester Free Press

CONTENTS

N O T I C E   T O   T H E   R E A D E R

Anthony Errington's autobiography appears here in print for the first time. In its original form, in manuscript, it lacks punctuation, omits and mis-spells words, and contains other difficulties. To make it easier to read and follow, I have edited the text, in ways described later. A GENERAL INTRODUCTION precedes the text of the AUTOBIOGRAPHY, and I have supplied a limited amount of explanatory information in the footnotes that accompany the text. I hope these editorial contributions, which sketch some of the historical context and explain certain technical points, will make the autobiography easily accessible to the general reader.

But some readers will require a more detailed explanation and analysis of the autobiography - perhaps because they want a fuller understanding of the references to contemporary coalmining, perhaps because they are interested in Anthony Errington as a manual worker during the British 'Industrial Revolution' or as a Roman Catholic, perhaps because they are local historians of Tyneside. The text is therefore followed by a number of ANALYTICAL COMMENTARIES. These summarise the information on such themes as Topography, Technology and Religion, and the longer Commentaries supply more detailed background and explanatory information than there is space for in the footnotes.

The present edition would not have been contemplated, worked on, or completed without the advice, support and assistance of many individuals. I refer the reader to the ACKNOWLEDGEMENTS at the end of the volume but express my indebtedness and gratitude at its start.

Finally, a detail. Desk-top composing of camera-ready copy allows as well as demands some departures from traditional printing conventions. Throughout this edition, terms of special significance are indicated by single inverted commas, and all quotations (as well as reported speech) by double inverted commas.

P.E.H.H.

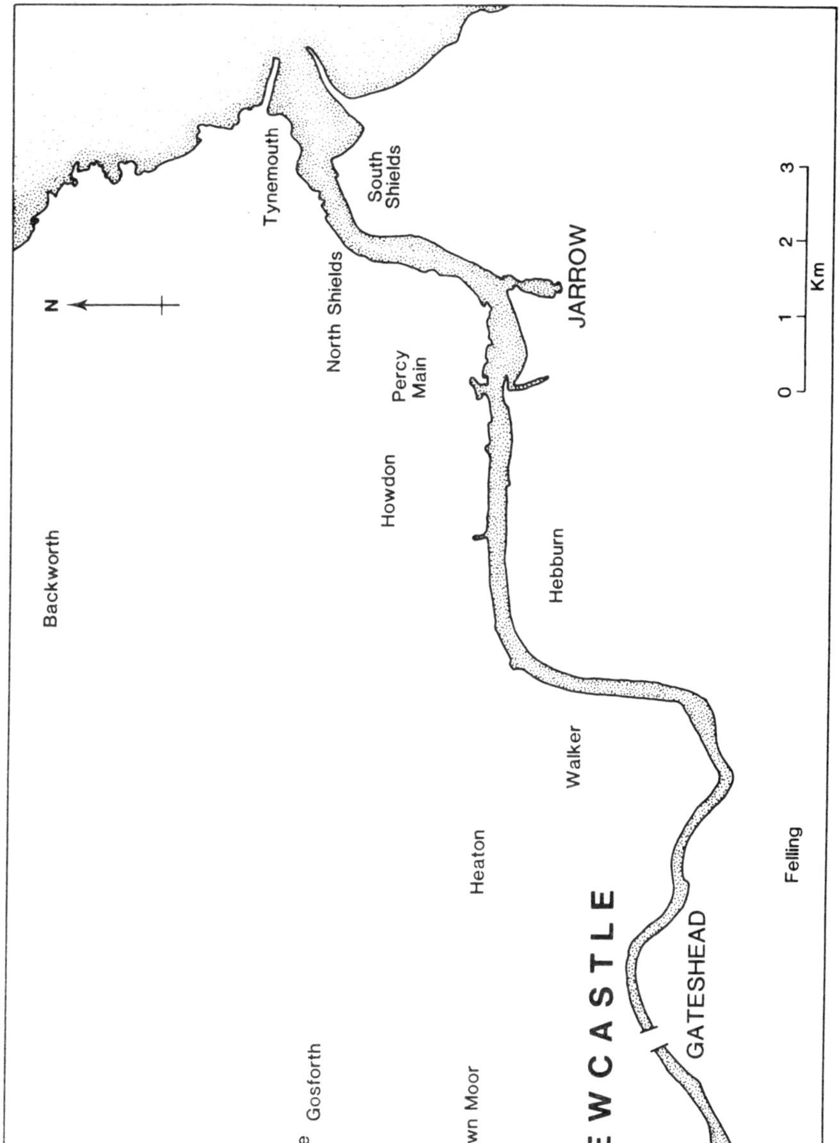

Backworth
N
Fawdon
Tynemouth
Coxlodge
Gosforth
North Shields
Kenton
South Shields
Howdon
Percy Main
Cowgate
Town Moor
Heaton
Fenham
Hebburn
Walker
JARROW
Benwell
NEWCASTLE
GATESHEAD
Felling
0 1 2 3
Km

LOWER TYNESIDE

Gateshead

N

GATESHEAD — SUNDERLAND — TURNPIKE

FELLING SHORE

Venture Pit

John Pit

LOW FELLING

Jarrow and South Shields

OLD SUNDERLAND ROAD, OR SPLIT CROW ROAD

Holly Hill Pit

William Pit

Durham

LOW HEWORTH

HIGH FELLING

Discovery Pit

Sunderland

Keppel Pit

0 Km 1

THE FELLING DISTRICT

GENERAL INTRODUCTION

Anthony Errington was born in 1778 and died, at the age of 69, in 1848. He probably began writing his autobiography in 1823, and the text we now have deals only with his life and career up to about 1825, when he was 47 years of age. He was born and died at Felling-on-Tyne, and he appears to have spent the whole of his working life, not only during the period of the autobiography but almost certainly thereafter too, within a few miles of his birthplace, among the collieries of the lower Tyne valley. His rank in society was that of a skilled tradesman, and his particular trade that of colliery waggon and waggonway wright.

Until his autobiography came to light, Anthony Errington was unknown to posterity. But he was a contemporary of three other natives of Tyneside who have found entries in the 'Dictionary of National Biography'. Although their individual careers led off in very different directions, all four men may well have rubbed shoulders in the lanes and alehouses of Newcastle-upon-Tyne. The lives of the three famous contemporaries, which in certain varying aspects parallel the life of A.E., or converge with it, can be usefully recalled, to provide a backcloth to the autobiography.

A.E. composed his autobiography in the 1820s. In the same decade, two of the contemporaries also wrote about their own earlier lives. Thomas Bewick, the engraver (1753-1828), wrote a memoir whose account of his childhood pranks and schooling in the rough countryside on the north bank of the River Tyne has much in common with A.E.'s less literate account of childhood pranks and schooling on the south bank of the Tyne.<1> Jonathan Martin (1782-1838), like A.E. marginally literate, and like him competitively battling in the artisan class of society, published in 1825 a history of his religious experiences and related social episodes, occurring partly on Tyneside, which foreshadowed his later notoriety as the incendiary of York Minster. Martin shared with A.E. a sincere but simple-minded belief in the manifest Providence of God, both as the mainspring of life and as the justification for an autobiography. One part of the immensely long title of Martin's book, "An Account of the Extraordinary Interpositions of Divine Providence on his behalf", would serve as a reasonably fair description of A.E.'s autobiography.<2>

The third contemporary shared with A.E. a different but equally important aspect of his life, the occupational environment of Tyneside coalmining and the transport of coal. George Stephenson (1781-1848), whose attainment of adequate literacy had been even more of a struggle than that of our author, during the period of the autobiography was living within a few miles of A.E. Although never an underground worker to the extent that A.E. regularly was, and although eventually promoting himself out of the class of skilled colliery workmen to which at one time both belonged, Stephenson laboured and schemed within the current technology of Tyneside coalmining, that is, among steam-pumps, ventilation devices, winding-engines, waggonways and inclined planes - items fundamental to many of the mining episodes recounted by A.E. At one time it seemed that Stephenson would achieve fame as the inventor of a miners' safety lamp. But instead his name is linked with the development of an effective steam locomotive, a machine designed to run on those waggonway rails whose laying and maintenance provided A.E. with his daily employment, as a waggon and waggonway wright.<3> Without A.E. and his fellow early railwaymen, there would have been no Stephenson locomotives, hence - perhaps - no world-wide public railways.

Waggonways and waggonway wrights

Coal was first mined along the Tyne in the Middle Ages, and by the end of that era large quantities were being exported by sea to London and to other ports in the coal-less southern parts of England. During the sixteenth century, the mines on the Tyne (together with those on the Wear, the second riverain inlet to the coalfield) greatly increased their output, to meet an enlarged demand for coal, locally and in the south, both for domestic and for industrial purposes. A bulky commodity, coal could be profitably transported in an age of poor roads only by water, hence the mines of the Tyne were originally dug at the river's edge. But by the seventeenth century these mines were exhausted and newer workings were started, further from the river. It was to carry the coal from these inland mines to the ships in the river that the earliest waggonways were laid.

At first the double-railed waggonway was built almost entirely of wood. But in the later decades of the eighteenth century, iron rails were introduced. At first these were of cast iron, but eventually of drawn iron. Meanwhile, mines were being opened further and further back from the river. By the time Anthony Errington became a waggonway wright, in 1792, some of the Tyne waggonways were ten miles long, and the total length of waggonways running to the river was well over one hundred miles. The further extension of the waggonways was limited only by the lack of economic traction power. The waggons were pulled by horses: only in a few steep places was there any mechanical traction - it took the form of a device that applied the gravity power of the descending full waggons to pull up the empty ones.

The Tyne collieries employed waggonways not only on the surface but also underground. The introduction of waggonways into the subterranean workings went on steadily throughout the eighteenth century, and by the end of that century there were as many miles of way below ground as above. Since the underground ways had regularly to be either lengthened or shortened as the working proceeded, and since they required even more maintenance than the surface ways, colliery waggonway wrights spent much of their time underground.

A.E. was taught his trade by his father, Robert Errington. Born the son of a farmer in a Northumbrian village about a dozen miles from the Tyneside colliery district, Robert followed his family when they moved south to a farm in the midst of the Wearside collieries. Around 1750, the father retired from the farm, and young Robert then went - as his son Anthony relates - "to the woods to work with Axe and Saw". Having learned the art of woodworking from his apprenticeship among the trees, Robert next turned to the more specialised trade of making waggons and waggonway. He began as a waggon and waggonway wright around 1755, and continued to make a living by this trade for 45 years, retiring in 1799 at the age of 67. In the same year that Robert Errington retired, his youngest son, Anthony, completed his apprenticeship to the same trade. Anthony worked as a waggonway wright from 1799, certainly until the mid 1820s, probably until the late 1830s, and he speaks in his autobiography of a brother who was apparently also a waggonway wright, and a son, Anthony Fenwick, who in 1818 was employed in the trade of his father and grandfather. Thus, for not less than 70 years there were Erringtons working as waggon and waggonway wrights in the collieries of Tyneside.

The years in which the Erringtons laboured on the waggonways were momentous years for the North East coalfield, as they were, partly in consequence, for industrial Britain. The period saw a rapid increase in the demand for coal, resulting on Tyneside in the multiplication of collieries and the intensification of coalworking; while the technical difficulties encountered - the greater distance from the river of the new collieries, the more persistent presence of water and gas in the newer, deeper seams - together with the pressure of vigorous economic competition, led to the introduction of a number of improvements in the techniques of coal production.

Little could be done to alter the process of coal-getting at the face, but the arrangements for the transport of the coal, from the face to the shaft, up the shaft, and from the shaft to the ships, were greatly modified and improved. After 1750, the use of waggonways underground, with horses or ponies to pull the waggons, became more general. From the 1790s, cast iron rails, and from the 1810s, drawn iron rails, began to replace the earlier wooden ones. From the 1780s onwards, Watt's rotary steam engine gradually replaced the horse-gin, water-wheel, or Newcomen engine employed previously to wind coal up the shaft (but the Newcomen steam pump survived in large numbers until much later). By 1800, inclined-plane haulage systems using gravity power were being incorporated in the waggonways wherever practicable, even underground. Finally, between 1812 and 1815, experiments with steam traction were conducted on the surface waggonways of four Tyne collieries, as follows: (1) at Coxlodge Colliery in 1813 by the owners, the Brandlings, using a locomotive built by John Blenkinsop, the viewer of the colliery on the Brandlings' Yorkshire estate at Middleton, near Leeds, where a similar locomotive had already been operating; (2) at Wylam colliery in 1812-1813 by William Hedley, the viewer; (3) at Heaton Colliery, also in 1812-1813, by Edward Chapman, a local civil and mechanical engineer; (4) at Killingworth Colliery, in 1814, by the enginewright, George Stephenson. As a result of these experiments, by 1823, the year in which A. E. probably began to write his autobiography, many of the Tyne waggonways were carrying steam locomotives; and while the autobiography was being written, the first public railways in Britain were opened.

Like Stephenson, the locomotive builders Blenkinsop and Hedley were almost exact contemporaries of the waggonway wright Anthony Errington, and all four were born at places within ten miles of Newcastle. Blenkinsop, before he was sent by the Brandlings to manage their Yorkshire colliery, was taught his trade by his cousin, Thomas Barnes, the viewer of Felling Colliery, and hence may have been personally known to A.E., who also worked at Felling under Barnes. Hedley built his locomotive at a colliery twelve miles from the one in which in that year A.E. was working; while Chapman at Heaton was only three miles away, Stephenson at Killingworth only four miles. The waggonways on which these early locomotives were tested were similar to those built by the Erringtons. As it happens, A.E. never once mentions steam locomotives (although he does once mention a steam boat). But it was waggonway wrights and colliery engineers who together brought the Railway Age into being. For coals on rails preceded, in Britain, all else on rails.

Underground dangers

Colliery waggonway wrights spent much of their time underground, and hence shared with the other colliery underground workers the vicissitudes of contemporary working life below the surface. A large part of A.E.'s autobiography is concerned with his adventures among the perils below.

His lifetime coincided with a dramatic period in the history of the struggle of the Tyne colliers with the natural forces of their underground world. In the later eighteenth century, thanks to the efficacy of the improved Newcomen steam pump, the colliers were able to work in deeper and wetter seams. But although they had conquered water, they found themselves menaced by another product of the depths - a gas which escaped from the coal, and which exploded when it came in contact with the flame of the candle by whose light the collier worked. Worse then followed: the explosion of this first gas generated a second gas which choked all human beings in its path as it was drawn silently through the airways of the mine.

Fire-damp (as the first gas, methane, was called locally) had been met in small quantities in the upper seams, and explosions with loss of life had occurred ever since coal was first worked in the Tyne basin. But these were infrequent and petty catastrophes compared with the major disasters that occurred in colliery after colliery in the last decade of the eighteenth century and the first fifteen years of the nineteenth century. Several hundred men and boys died in these years as the result of explosions. Adding the deaths from other forms of mining accident - inundations, shaft-accidents, roof-falls, etc - probably about one in ten of the colliers on the Tyne was killed each decade.

A.E. relates many narrow escapes from death underground, but refers only casually to what he might well have considered - in his own terminology - one of the most fortunate interventions of Providence on his behalf. In 1811 he suddenly decided to leave Felling Colliery, although he had worked there for nearly twenty years and his father had worked there for thirty years before that. On the 25th May, 1812, gas in Felling Colliery took fire and the mine exploded. Both shifts of men were below, 128 men in all, and 92 were killed. According to A.E., an old man had once prophesied to him that disaster would strike Felling. But since, by A.E.'s own dating of the episode, the prophecy was made about fifteen years before the disaster, it was not this which induced him to leave the colliery.

The Felling explosion had important consequences. Deeply disturbed by the distress caused, the incumbent of the parish, the Reverend John Hodgson, published a pamphlet describing the disaster and calling for measures to prevent further such explosions. Shortly afterwards, a 'Society for Preventing Accidents in Coal-Mines' was formed, whose leading members included Hodgson, and also John Buddle, the most distinguished mining engineer of the day. Buddle was then in charge of Percy Main colliery, to which A.E. had recently moved, and the autobiography tells how Buddle formed a branch of the new society among the miners at Percy Main. Buddle also wrote a pamphlet for the society on the ventilation of mines. The society achieved its greatest success when it approached Sir Humphrey Davy, the distinguished chemist, in the hope that he would discover a chemical method of neutralizing mine gases. Instead, Davy used gauze to produce, in 1815, the first practical mining safety-lamp.

The early safety-lamps were unpopular with the colliers because, although when used correctly they were safe to employ where there was gas, they gave a wretchedly poor light. They were therefore used only where the risks of explosion were greatest: in consequence, they were sometimes used too sparingly, with the result that explosions still occurred. Nevertheless, from the time that the safety-lamp was introduced, the mortality rate from explosions steadily declined. A.E.'s only reference to the safety-lamp is when he was exploring Tynemouth Castle in about 1820. He took a safety-lamp to a shop to be lit, and the shopkeeper described it as an "invincible lamp".

Economic and social discontent

These were stirring years for the colliers of the Tyne, and not only because of the struggle for underground safety. On the surface, the colliers also had their moments of struggle, as they engaged in spasmodic movements of communal economic discontent and militancy. A pamphlet of 1792 called the coalminers a "very turbulent set of men", and this was a reputation they had earned because of their many strikes and acts of disorder throughout the eighteenth century. The legend that before the repeal of the Combination Acts in 1824 the manual workers of Britain were cowed and submissive to their masters dies hard; and little that is scholarly has yet been written about the movements of economic discontent among the Tyne coalminers during the period of A. E.'s autobiography.

One year before Robert Errington settled as a waggonway-wright at Felling, all the colliers on the Tyne and Wear came out on strike. There was probably another general strike in 1792, the year that A.E. became a waggonway-wright, but if there was, the local newspapers failed to report it. The French Wars began, and many colliers joined the Army and Navy. By 1800, there was an acute shortage of labour in the coalfield, and the colliers seized the opportunity to exert pressure on their masters for higher wages. At this period, the colliers were hired on an annual contract which was renewed on the same date throughout the coalfield. They therefore formed a grand secret combination, and at hiring time refused to be engaged at any colliery unless the owner paid higher wages and a substantial hiring inducement. So successful was this combination at each hiring in the years 1800 to 1804 that the hiring inducement, which had been about half a guinea in the 1790s, rose in 1804 to between 12 and 20 guineas, the last sum being almost half the amount of a collier's annual earnings. In 1805, however, the coalowners in turn combined, refused to meet the demands of the colliers, and after a short stoppage, won the day.<4> The colliers' combination did not come to life again until 1810, when there was once more a general strike in the coalfield. Between 1800 and 1815, the colliers enjoyed a measure of prosperity and displayed the militancy of economic confidence. But the end of the war called a pause in the expansion of British industry, and the return of demobilised soldiers and sailors to the collieries gave the coalowners a surplus of labour. In these changed circumstances, the colliers were unable to exercise their economic muscle as actively. Little is heard about combinations, and there was no further general strike until 1825.<5>

It is difficult to assess precisely the social significance of the economic unrest of this period. Some writers imply that the workers suffered hourly from a sense of grievance, and that the dominating interest in their lives was the economic struggle and its political connotations. But there were several movements among the working classes which seem to have been at least as popular as the movements for economic or political advancement. And a large proportion of the workers had probably only the most spasmodic connections with any movement. Certainly political radicalism made little headway on Tyneside during the period of the autobiography, not even in the mild form of democratic reformism. And not even among the colliers, despite their history of cyclical economic militancy.

In his autobiography, A.E. makes not a single reference to the strikes and combinations of the colliers - nor does he mention the strikes of other local workmen, such as the keelmen (who were, if anything, more strike-prone than the colliers). The only economic unrest he mentions is his own when he thinks he is being underpaid, and he puts this right by moving to another colliery where he is better appreciated. As a waggonway wright he appears to have had no difficulty in finding jobs, even after the war; and at the collieries he experienced more personal contact than the average worker with the managerial staff - deputies (underground foremen), overmen (shift foremen), viewers (managers), coalowners' agents - and even with individual coalowners.

A.E.'s autobiography must therefore come as somewhat of a disappointment to historians of 'working class consciousness' and to the more ideological exponents of 'history from below'. Indeed it raises doubts about the validity of the former concept, since A.E. is an individual more in line with old-fashioned views about the working man of the period than with the views of post-1960s revisionism. Deferential to his superiors, uncritical of the economic regime, individualistic and striving for personal advantage, a family man, decent and reasonably respectable in language and behaviour, patriotic and Christian - A.E. is a proto-Victorian would-be bourgeois, and his autobiography has more than a smack of Samuel Smiles. It probably does have a 'hidden agenda' but one that is far removed from social revolt.

This said, the response must be conceded: A.E. is not a typical working man, for quite apart from his religious idiosyncrasy, he is not even a typical collier. Although he shares the perils of the colliers underground, in terms of social status A.E. is a skilled tradesman and not a normal coalminer. Yet tradesmen sometimes had their own associations, and when at one point A.E. refers to a man as a "Brother trade man" the possibility arises that waggonway wrights had their own 'brothering' or trade combination. But nothing is known about such an association - which if it existed may have been only a friendly society or even merely a convivial club - and this is the nearest A.E. ever gets to being other than an independent operator.

A waggonway-wright was not normally engaged at an annual hiring as the mass of colliers were, but was taken on when work was pending or when a replacement was needed. During A.E.'s lifetime new collieries, and within them new pits, were regularly being opened, requiring new surface and underground ways to be laid and new waggons to be built. Moreover, every working pit needed the services of several waggonway wrights, to extend the undergound way into new workings or withdraw it from retreating ones, and to maintain the existing surface and underground way and waggons. The waggonway wright worked partly on a piece-contract, undertaking to build a particular length of way or a particular number of waggons, and to maintain these for a given period. He was paid fortnightly like the colliers, but his net earnings were subject to a cost-bonus system, whereby part of the cost of the materials in the waggons or the way he was building came out of his own pocket, in return for which he earned a bonus as long as he maintained them. The average earnings of a waggonway wright appear to have been , at best, only a little higher than those of a collier. A.E. mentions receiving around £1 a week. Colliers' wages probably averaged, in the first decade of the nineteenth century between £1 and 25s, and in the second decade between 15s and £1. It was the regularity of employment and the fact that at work he was largely his own master, rather than the economic advantage, that gave the waggonway wright a higher social status than the ordinary collier.

Recreations

Tyneside workers had many leisure-time activities. Politics was not one of the popular activities, and neither in the 1790s nor in the decade after 1815 did the propaganda of the democratic reformers make much impression in the Tyne valley. A national movement which was much more successful there, at least for a time, was that of Methodist religious enthusiasm. Weslcy himself had preached in the colliery villages around Newcastle, and during the French Wars the influence of Methodism was considerably strengthened. Soon there was a Methodist chapel in every second mining village, and the meetings in these were attended, more or less regularly, by a significant minority of the colliers. How Methodism affected A.E. will be described later.

But most of the workers found other ways of spending their leisure time. The alehouse occupied a fair amount of that time. The temperance movement was only just emerging and the practice of drinking alcoholic beverages at every point of social intercourse was still respectable. A great many of A.E.'s adventures begin or end in the alehouse.

Walking (albeit sometimes from alehouse to alehouse) and outdoor sports were other favourite activities. Fortunately the Tyneside worker did not have to travel far to find open spaces, green fields, and fresh air. This can be seen from the following description of the lower Tyne valley in A.E.'s day.

Tyneside around 1800

Then as now, the centre of population on the lower Tyne was Newcastle, eight miles inland, ten miles up the winding river. A town maintained by commercial and administrative activity, it stood at the north end of the only bridge across the lower river. With Gateshead, its small sister-town at the other end of the bridge, it accounted in 1800 for a population of some 40,000. Newcastle had already expanded beyond its medieval city walls and was about to take in several small mining and agricultural hamlets. Gateshead was to expand in the same way during A.E.'s lifetime.

Down-river from Newcastle and Gateshead, another 40,000 persons lived in riparian towns, villages, and hamlets. Half lived in the fishing and general ports of North and South Shields, near the river mouth, the rest in a score of small settlements, many of which owed their existence to nearby collieries. On the south bank, between South Shields and Gateshead the main settlements were Jarrow, Hebburn, Heworth and Felling: on the north bank, between North Shields and Newcastle, the main settlements were Percy Main (in 1800 newly founded), Howdon, Willington, Wallsend, Walker and Byker. Almost all of these localities are mentioned in the autobiography.

Coal was the principal industry of the whole district. The river was generally crowded with vessels - especially coal ships lying off, and the little boats, or keels, which carried the coal from the terminus of each waggonway out to the ships. The coal trade was further served by ship-building and ship-repairing establishments, by ropeworks, and by the iron and chain manufactory at Gateshead. As yet there were few other large-scale industries on Tyneside, but the declining manufacture of salt was soon to be replaced by the rising manufacture of chemicals, located not least at Felling.

In the villages and hamlets nearest the river lived sailors, keelmen and the workmen in the ancillary industries; in the villages and hamlets further back, and near the mines, lived the colliers. Most of the colliery settlements contained only two or three streets of houses: some of them had been founded since A.E.'s birth. Between these settlements and behind them, the countryside was often bare. North of the Tyne stretched large expanses of moor, south the land rose quickly to fell; both moor and fell were as yet little cultivated and were occupied only by occasional cottages, often of squatters. But A.E. lived to see the frontier of intensive agriculture advance in both directions towards the Tyne. For although the deep countryside barely appears in A.E.'s story, the farmland areas of Northumberland and Durham had already experienced a measure of technological revolution, agricultural intensification, and marginal expansion. And as a result, they were still managing to feed the growing population of Tyneside, with rye and wheat for its bread, oatmeal for its gruel, and barley for its beer, as well as with large quantities of meat and cheese.

Where Anthony Errington worked and lived

A.E. was born at Felling, a village, or more correctly, a group of hamlets to the east of Gateshead. Felling Shore, Low Felling and High Felling occupied the positions suggested by their names on the slope of the fell rising from the river. They were too small to have much social life of their own, and the Felling man who sought a night's conviviality, or the Felling woman who wished to restock the larder, generally tramped a long mile into Gateshead or another mile across the bridge into Newcastle. The hamlets of Felling had developed mainly since the opening, in the late 1770s, of Felling Colliery, the first large-scale mine in the locality. Like many other settlements which had appeared since the ecclesiastical partitioning of England in the mid Middle Ages, they were not part of the parish with which their social ties were closest, that is, Gateshead. Instead they formed part of the parish of Jarrow, a town four miles down-river, once a monastic settlement and the home of the venerable Bede, but by 1800 no more than a small mining village. Jarrow parish being so large, it was split into two; and Felling was in a chapelry which took its name, Heworth, from the village next to Felling in which the chapel was actually situated.

The registers of the chapel of Heworth contain the records of the marriage of A.E.'s parents, of the births of some of their children, including Anthony, of Anthony's first marriage, of the birth of one of his children, and of the deaths and burials of several of the family.<6> But A.E. was not an adherent of the Established Church. He was a Roman Catholic and a loyal son of his church, and he therefore probably never attended Heworth chapel except when he was married there (as the law demanded). But he went to school at Heworth, and one of his adventures as a school boy led to his being nearly buried alive in Heworth churchyard. The countryside around Felling and Heworth was a paradise for schoolboys - above the villages were bare fells; between the villages were fields and woods, with occasional pits and waggonways; below the villages were bustling quaysides and the busy river.

At the age of thirteen, A.E. began work in Felling Colliery, and after being bound apprentice to his father, worked there for seventeen years. He would probably have remained there even longer, as his father had done, if his home had not been broken up by the death of his wife. Felling Colliery at that time was a mine of only moderate distinction. Owned by a prosperous local family of the landowning squirearchy, the Brandlings (after whom it was sometimes known as Brandlings' Main Colliery), its several pits were then working the High Main seam. Not until 1803 were sinkings made to the deeper Low Main seam, and A.E. left the colliery shortly after the mine ceased working the High Main, in 1810. The transference of all activities to the Low Main brought the colliery a certain notoriety, first, because of the explosion of 1812, and secondly, because the abandonment of pumping in the High Main led within a few years to the drowning out of several collieries still working that seam lower down the coal basin.

From Felling, A.E. moved in 1811 to Percy Main Colliery, a few miles down-river, but on the other side, on the north bank. While Felling was in the County Palatine of Durham, Percy Main was in the County of Northumberland: the Dukes of Northumberland, the Percys, owned the land and gave their name to the colliery sunk there in 1799. Percy Main - both the colliery and the village which grew up round the colliery - was only a dozen years old when A.E. arrived. Here, as elsewhere, he almost certainly lived in a colliery house, a house owned by the mine and provided for him, as long as he worked there, against a fairly notional rent. The infant settlement of Percy Main must have had little in it to occupy a single man, and A.E. most probably spent much of his leisure time either in nearby North Shields, or back in Felling - boatmen carried passengers across the river - where his aged father still lived and where his younger children were being brought up. In his loneliness A.E. came close to acquiring a second wife while at Percy Main. The colliery worked the High Main coal in two pits, the Percy, opened in 1799, and the Howdon, opened in 1804. The chief viewer (manager) of the colliery was John Buddle of Wallsend, already the most eminent viewer in the coalfield, who was to become a friendly acquaintance of A.E.

In 1814, A.E. left Percy Main and went to Fawdon Colliery, a couple of miles NW of Newcastle. Fawdon was an even newer colliery than Percy Main, having been opened in 1810, and the village of Fawdon contained only a few dozen houses, which is perhaps why A.E. lived in the next village, Kenton. The Fawdon workmen were in the habit of walking into Newcastle, across the intervening Town Moor; and on this moor, after a visit to a Newcastle alehouse, A.E. twice saw a vision relating to the disastrous inundation of the nearby Heaton Colliery.

He moved again, probably in 1816, to resume work under his former employers, the Brandlings. As well as their estate at Felling, the family owned land immediately north of Newcastle, at Gosforth; and there they worked a small colliery. To this they added in 1809 a new mine at neighbouring Coxlodge; and eventually in 1825 opened a new and deep mine at Gosforth. Coxlodge Colliery waggonway was the scene of the experiment with the Blenkinsop locomotive in 1813. A little later, the head of the Brandling family, having become acquainted with George Stephenson, was his prime backer in his claim to have been the first to invent a successful safety lamp. A.E. worked first at Gosforth, and then at Coxlodge. He does not tell us where he lived, but wherever it was in this locality, he was again within walking distance of Newcastle.

In 1818, A.E. moved once more, his fourth move in seven years. This time he moved further back from the Tyne than he had ever done before. Backworth Colliery, his new station, was four miles north of Percy Main, in part of the coalfield previously unworked, and the colliery was only opened in the year A.E. joined it. A group of partners owned the colliery, one of whom was the Mr Taylor mentioned in the autobiography, while another was John Buddle, the mining engineer. The waggonway from Backworth to the Tyne was unusually long, and in 1821 a stationary steam engine was introduced to haul the waggons on ropes - a cautious stage towards the employment of the steam locomotives that were already at work on the neighbouring Killingworth waggonway. Again, A.E. does not state where he lived, but in 1818 he married again, and his first child by his second wife was born at Backworth.

Despite this renewal of marital domesticity, at an unstated date which may have been 1822, A. E. moved yet again, this time to Walker, a colliery sixty years old, near the Tyne and only one mile east of Newcastle. From Walker, Anthony could look across the Tyne and see on the south bank the hamlets of Felling where his industrial pilgrimage had begun. It may have been at Walker that he began to write his autobiography, and it certainly breaks off with his arrival at Walker.

It is possible that he did not stay long there. Evidence other than his own writing suggests that around 1825 he may have returned to Backworth or to a neighbouring colliery. In 1831 he was most probably living at Felling, since his last child by his second wife was born there; and presumably he had returned to work at Felling Colliery, where he began as a waggonway wright forty years earlier. How long he worked before he finally retired is not known, nor is it known whether he continued to live at Felling throughout the period up to his death there in 1848. But it is plausible that Felling was the scene of the last two decades of his life as it had been for the first three.

INTRODUCTION

The reason for the autobiography

Assuming that the first line of the text written below his name carries this significance, 1823 was the year in which A.E. put pen to paper and began to write his autobiography. It was an unusual exercise for a marginally literate man. What induced him to do it? The most likely explanation is that the autobiography was intended to demonstrate the respectability of his religious allegiance. As David Vincent has recently stressed, A.E. wrote "to inform my famely and the world", that is, as well as recounting his adventures for the entertainment and edification of his descendants, A.E. had some notion of making public what he wrote and thereby influencing others.<7>

A. E. was a Roman Catholic. It is one of the several odd features of the autobiography that his family's religious persuasion is only twice directly mentioned. Nothing in his terminology when speaking of the Christian religion, or when repeating prayers, indicates that he was other than an English protestant of the age. In a later period (almost up to the present day), such eirenic conduct on the part of a Roman Catholic would seem strange. But the spirit and the circumstances of the Catholic church in England were very different in the period of the autobiography. Although no longer actively persecuted, the church's relatively few adherents found it wise to lie low, and not improper to assimilate their way of life, and even their spirituality, to those of the protestant majority in the country, to as great an extent as was theologically plausible. Paucity of priests and chapels made it difficult for the laity to receive rigorous instruction and close clerical supervision. Hence, left partly to their own devices, they were content to pass on the tradition of the Old Faith to their children, without emphasising publicly those aspects of it that were likely to aggravate their protestant neighbours.

A. E.'s name appears in the incomplete registers of the Roman Catholic chapel at Newcastle, the chapel nearest to his home at Felling. One of his adventures begins with his setting out on a Sunday morning to attend this chapel at Newcastle. Four of his children, including a natural child, were baptised there. He further appears in the register on several occasions as a godfather.

The references to the family faith in the autobiography are casual and unassertive. In one anecdote A.E. and his father are referred to as "papists", and at another point A.E. notes that he sought advice on the subject of ghosts from the "Catholic preast at Newcastle". Slight though these references are, in pointing to two aspects of his religious belief they provide a clue to much of A.E.'s mentality, and help to explain some of the more puzzling episodes in his life.

A.E. recounts how, as a child, he took part in scenes of dairy magic, when he compelled butter to set. The words of the spell he used were from the protestant terminology, and it was not unknown for protestants to practise dairy magic. But the gesture he admitted making on one occasion, and probably made on all three occasions, "the sign of the Cross" - presumably crossing himself - was a piece of non-protestant ritual. The Old Faith and the Old Magic were no doubt to some extent linked and confused in the minds of the protestant peasantry: strange gestures, wonder-working objects, mysterious sacraments continued to give the impression that all papists were in closer touch with the supernatural world than were simple protestants. Hence, when the butter refused to set, the Catholic Erringtons were sent for.

A.E. was a very superstitious man. The degree of superstitious credulity which appears in his autobiography was almost certainly not typical of the Tyneside workman of this period. Heavy drinking and the sudden disasters of the mine gave the colliers some excuse for seeing visions and noting omens, yet evidence from other sources suggests that they were on the whole rather a stolid and unsuperstitious set of men. There were exceptions and we meet some of them in A.E.'s account. But A.E. suggests many times that he has been specially favoured and has been vouchsafed visions and omens denied to other men. Thus, to return to his religious faith, he suggests in one very significant episode that Roman Catholics are abnormally sensitive to the supernatural. The episode is that of the first vision on the Town Moor. A.E. and a friend specifically stated to be a Catholic are walking in company with two other men who are arguing about points of Scripture, that is, they are protestants.

The vision appears to the two Catholics: the protestants, blinded by their bibliolatry, see nothing.

In A.E.'s earlier days, the Roman Catholic faith earned its adherents little public respect. In 1825, a 'Catholic Religious Defence Society' was formed at Newcastle, "in order to stem the torrent of calumny, misrepresentation and abuse, which is so lavishly poured forth by certain bodies of men, styling themselves 'Religious Tract Societies', 'Continental Societies', 'Irish Evangelical Societies', 'Gospel Trust Societies', etc. etc. against the Catholics of the United Kingdom". The misrepresentations and abuse complained of were nothing new; but the protest and aggressive reaction were. The Catholic Revival was under way, for after the French Wars English Roman Catholics began to demand more public respect for their faith. The Catholic Religious Defence Society at Newcastle aimed to publish and distribute gratuitously works "defending our principles"; and it is likely that A.E. had the same general object in mind when he began his autobiography, only two years before the Society was founded.

The most active, or at least militant, protestant sect on Tyneside in A.E.'s day was 'the Methodists', and it is not difficult to imagine how a loyal but rather simple-minded Catholic would fare when forced to work alongside these protestant enthusiasts. One of the most popular beliefs of the Methodists of the period was a strained interpretation of the Christian doctrine of Divine Providence. The early **Methodist Magazine** filled many of its pages with anecdotes on the following lines - two men descending a pit shaft, one a Methodist singing hymns, the other a blasphemer singing bawdy songs; the rope surges, the blasphemer falls and is killed, the Methodist is saved; yet again "the Providence of God is illustrated". The crudity of some of the anecdotes, and hence of the underlying belief, can hardly be exaggerated; but obviously such anecdotes were very well received by sections of the Methodist rank and file. The phrase 'Providence of God illustrated' was for a time so much heard that one non-Methodist in Newcastle was sufficiently galled to issue a handbill which read: "Stage Coach Overturns. Four Methodist Preachers Travelling to a Conference killed. Is this the Providence of God illustrated?"

INTRODUCTION

A.E.'s autobiography could in part pass for that of a Methodist of his day: the first paragraph speaks of "the hand of Divine Providence being over me in all my actions", and later he writes - "I was possessed with a lively Faith in the Providence of God over Mee, and I cared not what the wurld said on me." As already noted, the eccentric Methodist, Jonathan Martin, wrote in similar terms; and so did a humbler soul, John Thompson, class leader of the Methodist Society at High Felling, who was killed in the 1812 Felling disaster and whose diary was instantly published by his co-religionists. This cosmic interpretation of the superficially contingent explains and partly justifies the apparently self-congratulatory nature of many of A.E.'s anecdotes. In recording his successful (sometimes astonishingly, or even unbelievably, successful) feats in life, A.E. is not praising himself but thanking Divine Providence. Thus, A.E. is out-Methodisting the Methodists: he is showing them, and the world, that Divine Providence is extended to the much-abused Roman Catholics, indeed that Catholics may sometimes receive more than an equal share of the divine attention.

A.E.'s attempt to win respect for his own faith would probably not have impressed many at the time, had it been known. But the modern reader may well manage some admiration for A.E.'s guileless loyalty to his creed, and feel that the record of his acts of tolerance and kindness to those of other persuasions does no discredit to its teaching. There is a warm human appeal in the brief episode of George Hawkes and his wife, drunk.

"That Evening ... he had got two mutch and Cowped [= upset] the Cart at the Robers Corner ... I had to set to and get them gathered Out of the gutter ... Shortly after, I met them again in Gateshead. He had just denied My Fauther of carrieing his Market Poak [= sack] in his Cart to Felling. I told him One good turn deserved another. But they said that was not their wish, to do any good turn to a papist ... 3 weeks after, I found them in the same ditch at Robus Corner. I Stood and Looked. They was no worse but boath drunk and the Limer of the cart broke. I gethered them up for the Second time. She said, 'I thought a papist would not have done so to a protestint'. I left them to go home."

INTRODUCTION

Death and character

Although the autobiography breaks off around 1825, Anthony Errington lived on until 1848. His death-certificate tells us little more about him. He was a waggon wright, aged 69, living at High Felling: the cause of death was "Disease of lungs 5 months Certified": the informant was his youngest child, Mary Ann Errington, aged 17, who was present at the death. "Taking this with me, that their is a Just God to judge me in life and after death, that I may meet the Just in that Blesed manshon of Bliss wheare the Just rain for all Eternety with their God, through the redempshion of Our lord and Saviour Jesus Christ. This is my hope of reward..."

From the evidence of the autobiography, it is easy to see what character A.E. envisaged himself as having, easy to detect certain aspects of character that probably escaped his self-observation, but not easy to draw up a rounded picture of the man. Loyalty was a marked feature of his mentality. Family loyalty prevented him from uttering a word of criticism of any relative. Loyalty to his employers and industrial superiors, and a degree of deference to both, were only shaken when he felt that employers and viewers did not respect the bond. Religious loyalty was also no doubt very strong, although it tends to be concealed in the autobiography. Loyalty to the nation was expressed in patriotic sentiments (yet A.E. manages to be as unspecific about the post-1793 wars with France as his distinguished literary contemporary, Jane Austen). Only class loyalty is absent - unless we make something of the episode when he is supported by the rank and file, the hewers, against the foremen, the overmen, who try to shout him down at a safety meeting. But loyalty, like the sincerity, charity and honesty which the autobiography aims to highlight, may look differently from without than from within. A.E. was not a sophisticated individual, or at least his literary weaknesses inhibited any sophisticated expression, and therefore not enough critical awareness appears in the autobiography to reveal much to us in respect of what others thought of the man or what lay behind the self-portrait. Readers of the autobiography may therefore legitimately hold a range of opinions about its author.

But what certain points in the account suggest to me is that A.E. was at times thought of as a rather simple soul, easily made a fool of and perhaps to some extent a figure of fun. It is difficult to be sure whether he was in fact just mildly eccentric, a mood quite conceivably induced by his being set apart as a Catholic, or whether he was truly rather simple-minded. At times he seems shrewd and even intelligent, and on balance I incline to the view that he was no fool but was unable in his life and writing to conquer the disabilities of inadequate education. If instead he was a very simple soul gallantly determined to express his faith, then my judgement is perhaps patronizing; and I can only apologise to him by making a further incautious judgement. Despite the danger of relying on inadequate and subjective evidence, I find A.E. a nice man, a 'canny soul' in the best sense of the Newcastle expression. It is possible to laugh at him and yet respect him, even perhaps admire him. I have enjoyed meeting Anthony Errington and I hope the reader will do so too.

The manuscript and its editing

Anthony Errington's manuscript autobiography is in the Local Collection of Gateshead Borough Library. I describe the manuscript, its history, and my editorial procedures elsewhere in this volume, in detail. I believe the manuscript to be written in its author's own hand. It is almost entirely lacking in punctuation, and I have added this, as well as the additional material shown in square brackets. The curious spelling, often representing Tyneside pronunciation, has been generally retained. Dialect and technical terms are explained in the text and also in a Glossary.

Notes to the General Introduction

<1> Bewick grew up around the village of Ovingham, 12 miles west of A.E.'s childhood district, and just west of the main Tyne coalfield, although his father's farm did include a small coalmine. As an adult, Bewick lived in Newcastle from the 1780s, but the company he kept, and the alehouses he spent many evenings in, were of a higher social class than A.E.'s. His memoir, written mainly at Tynemouth, was only published in 1862.

<2> Martin grew up around Hexham, on the upper Tyne, had no formal schooling, and completed an apprenticeship as a tanner before becoming a seaman. Leaving the sea in 1810, he worked at his trade in south Durham, but his religious eccentricities, building on his Methodist attachment, led to his being confined in madhouses. In 1821, he escaped from Gateshead madhouse, as described triumphantly in his 1825 autobiography. An 1826 edition of the autobiography included three crude illustrations contributed by his brother William (1772-1851), parish-pump eccentric and inventor, who lived at Wallsend from 1813, and who sold or gave away in the streets of Tyneside copies of his own more than two hundred pamphlets and broadsides (and who therefore was almost certainly known to A.E., at least by reputation). William, who worked in the 1790s at a ropery at Howdon, near Wallsend (during the same decade A.E. and his father worked for a time at a rival ropery across the river), later involved himself in the mines safety movement, offering advice to the Wallsend colliery engineer, John Buddle - as A.E. also did. A third brother in this extraordinary family was the distinguished London painter, John Martin (1789-1854). If we can believe William's autobiographical fragments, in 1805 the three brothers paid a visit to Percy Main colliery - where A.E. was to work in 1811-1814 - and John used his burgeoning artistic talent to sketch out one of William's coalmine inventions. Later in life, John, who, as well as painting apocalyptic landscapes designed sewers for London, remembered his Tyneside coalmining connections and published a plan for preventing explosions in mines. On the brothers, see Thomas Balston, **The life of Jonathan Martin** (London, 1945).

<3> Stephenson was born at Wylam, west of Newcastle, married at Willington Quay, near Wallsend, and from 1805 worked at Killingworth, NE of Newcastle. Like William Martin (previous note), in his earlier life Stephenson worked on both a Perpetual Motion machine and a rival safety lamp to Davy's. There are many parallels between episodes in A.E.'s autobiography and episodes in Stephenson's childhood and Tyneside career, as recounted in Samuel Smiles, **The Lives of the Engineers**, vol.3 (London, 1862).

<4> I have described this period of remarkably successful 'combination' on the part of the North East coalminers, in P.E.H.Hair, 'The binding of the pitmen of the North East 1800-1809', **Durham University Journal**, 58, 1965, pp.1-13.

<5> While standard accounts concentrate on the occasional, spectacular coalfield stoppages, the work diaries of the colliery viewers modify the picture of sharp swings between workers' passiveness and militancy. They show that, due to the nature of coalmining, there were almost continuous negotiations between the viewers and groups of coalface workers over working prices and other matters, as influenced by continually changing working conditions, and that while there were regular confrontations and stoppages at this level, at the same time there was developing an overall, recognised system of industrial bargaining and compromise.

<6> A.E. himself was buried at Heworth, as were his parents and a sister.

<7> David Vincent, **Bread, knowledge and freedom: a study of nineteenth-century working class autobiography** (London, 1981), p.21. In the context of his perceptive analysis of other working class autobiographies, Vincent comments further that "Errington's limited command of the techniques of writing makes it particularly easy to hear the voice of the story-teller"; and suggests that "most of the memoirs have a central bearing on the formation of his character, but a proportion are included for no better reason than that their recollection amuses the writer and may entertain the reader". The extent to which the individuality of A.E.'s autobiography can be fitted into generalisations about working class autobiography is debatable, but Dr Vincent's book is commended to the reader who wishes to see the autobiography in wider perspective.

Recommended background reading

The best brief introduction to the history of the North East region in the period of the autobiography is in the first part of Norman McCord's **North East England: the region's development 1760-1960** (London, 1979). There is no good history of the Felling area. F.W.D.Manders' **A history of Gateshead** (Gateshead, 1973) is more rounded and adequate for the period than S.Middlebrook's **Newcastle upon Tyne: its growth and achievement** (Newcastle upon Tyne, 1950). But the best insight into the tone of A.E.'s locality in the period of the autobiography is provided by the relevant sections of the splendid early local histories, especially those by Hutchinson (1787), Brand (1789), and Mackenzie (1825,1827,1834) - for full titles of these works, see the Bibliography elsewhere in this volume.

On the general history of British coalmining in this period (when the North East was still Britain's major coalfield), the technological and economic aspects are comprehensively and authoritatively dealt with in Michael W. Flinn's **The history of the British coal industry, vol.2 1700-1830** (Oxford, 1984). However, on the former aspect this work does not entirely supersede a pioneering study, R.L.Galloway's **Annals of coal mining and the coal trade, First series** (London, 1898, reprinted 1971); and while the Flinn volume makes frequent use of the manuscript records now in the Northumberland County Record Office, there is scope for a more exhaustive analysis of these.

On the social history of the miners, a more critical and empathic approach than that in the somewhat dreary histories of ever-triumphant trade-unionism is provided in John Benson's **British coalminers in the nineteenth century** (Dublin, 1980) - albeit this work far outruns A.E.'s lifetime. Robert Colls' rambling essay in proletarian consciousness-raising, **The collier's rant** (London, 1977), discusses dismissively the North East 'popular' literature of the period. A 'dog in the night' aspect of the autobiography, A.E.'s lack of interest in current anti-Establishment politics, may be seen as part of a wider apathy, explained somewhat apologetically in H.T.Dickenson, **Radical politics in the North-East of England in the later eighteenth century** (Durham County Local History Society, 1979).

As an introduction to the work of a waggonway wright, see the careful monograph by M.J.T.Lewis, **Early wooden railways** (London, 1975). The most recent survey of Tyneside waggonways is in C.R.Warn's **Rails between Wear and Tyne** (Newcastle upon Tyne, 1985).

The only work to discuss A.E.'s autobiography to date is the study by David Vincent listed in note <7> above.

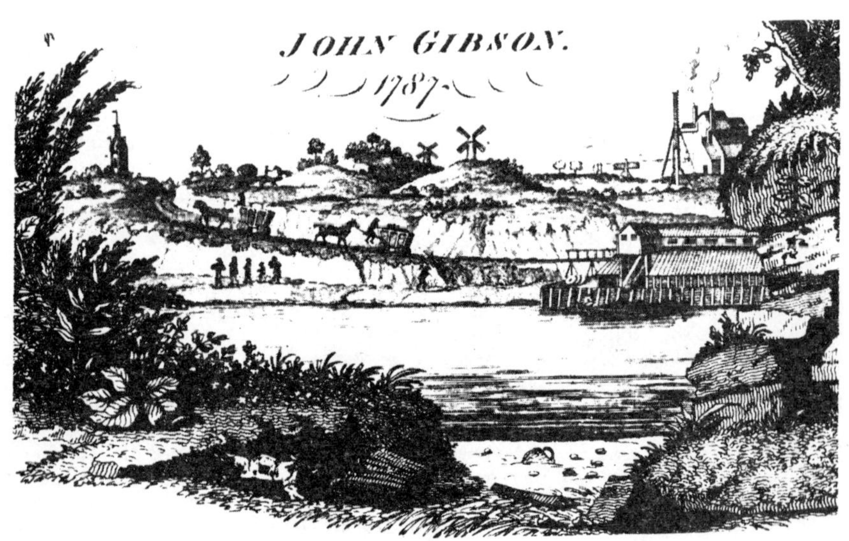

THE AUTOBIOGRAPHY OF ANTHONY ERRINGTON

Anthony Errington

the 21 of September in the year of our Lord 1823

The reason of my wrighting <1>

The reason of my wrighting the particulars of my life and Transactions are to inform my famely and the world. [And I say] that I wright this from pure motives of Justise and thouth [truth], and that [both] with and agains my self. I shall Explain every Transaction as Breaf as posable. Taking this with me, that their is a Just God to judge me in life and after death, that I may meet the Just in that Blesed manshon of Bliss wheare the Just rain for all Eternety with their God, through the redempshion of Our lord and Saviour Jesus Christ. This is my hope of reward, the hand of Divine Providence being Over me in all my actions, so that I have often repeated that [saying] of Pacient Joe, that all Things woork together for good - an Old Collier of Benton moore, the Poat.*

I shall devid my life to the Diferent stages, that of my Birth and famely, that of youth at school, that of Prentishop, that of Jurniman, that of Marrigee, that of Widdower, that of Marrige.**

---

* 'Patient Joe, or the Newcastle collier', who "was certain that all work'd together for good", was the exemplary hero of a tract in verse by Hannah More, published anonymously in 1795. Did A.E. recognise the Biblical source (Romans 8.28)? Joe "dwelt on the border of Newcastle town" which A.E. interprets as Benton Moor, NE of Newcastle; and he appears to think that the author of the verse was Joe himself.

** A.E. was married twice. The autobiography does not adhere strictly to the scheme here mentioned.

My farther and my mother <2>

My farther, Robt Errington, was Sun of William Errington of Nither Witten, Northumberland, Farmer at the Langlee, and [who] removed from their to Kiblesworth, near Lamsly, in Durham, wheare they farmed.* My grand farther being aged, they declined farming, [and] my farther going to the woods to woork [with] the Axe and Saw, which he Continued at for some years, and [later] hired himself to make waggons and waggonway, which business he continued for 45 years. And [then] took a small farm at Felling and Declined business as [a] waggon wright, and lived their to the age of 86.**

My Mother was born at Matfin Moore Houses and when very young [left] the famely. Her farther being dead and the step mother maried again, she had to make plase at 11 years of age, and was at sarvise at Powders Close, near Hebron quey, on the South side of the Rivir Tyne.*** When my farther and her married, they set up the first house at High Felling, in the Chepelry of Hueth [= Heworth] in the Parish of Jarrow,

---

* The register of the parish church of Netherwitton, a farming village 20 miles NNW of Newcastle, shows that Robert, son of William Errington, was born there in 1732. Langlee, or Longlee, was and is a farm near Netherwitton village; Kibblesworth and Lamesley, neighbouring localities about five miles south of Newcastle, were then agricultural villages. Many Erringtons were still living at Lamesley in the mid nineteenth century.

** Robert Errington (1732-1818) worked as a waggon wright from 1754 to 1799, perhaps from the start at Felling. In the eighteenth century Felling was a group of hamlets (High Felling, Low Felling, Felling Shore) situated on the east side of Gateshead. During A.E.'s lifetime these hamlets began to coalesce and to become to some extent industrial outliers of Gateshead, as it expanded towards them. Robert Errington lived at Felling from at least 1766, and Felling is the scene of much of his son's autobiography.

*** Matfen was and is a farming village located 15 miles almost due west of Newcastle. "Hebran" (Hebburn) Quay, on the south bank of the Tyne, five miles east of Gateshead, was at this date a hamlet.

County of Durham.* And their Dwelled and Brought up 4 Suns and 5 Doughters.** My sister Isable Died of the Warter in the Brain at 11 years of age. She was called after my mother, Isabella Errington - her maiden name being Carr. And she was a very very industrious dutiful mother amongst her family.

My farther, the very patron of Industry and honesty, was beloved by all, all ways ready to do good to any one. He loved to be in Sochiel Company, and all ways endevered to restore peace when any frocthan [= friction] took plase. He was the first that put two Convoys [= hand-levered brakes] upon the pit waggons, furst that Invented the Double Switch, first that made the Self acting Incline of waggonway in the Ann pit of Felling, first that made the waggons for the Old Faud Colliery, near Gateshead, first that laid the waggonway in the rope walk, [at] South Shore, for Chepmans patent lay, in the year of 1799.*** He Rode the first waggon of Coals from the An pit, Felling, to the Rivir Tine, high main coal. (Felling Colliery was begun by Charles Brandling Esq 15 of October 1777,

---

* "Hueth" (Heworth) was at this date a village, made up of two hamlets, Low or Nether Heworth and High Heworth, located immediately to the east of Felling; and Jarrow, once a renowned medieval monastic settlement, was a mining village further east. Felling, although four miles west of Jarrow, was included in Jarrow parish, in its chapelry of Heworth.

** According to the register of Heworth Chapel, Robert Errington and Isabella Carr were married there in 1766. Because they were Roman Catholics, their children were not baptised in the Anglican church, but the births of four sons - in 1772, 1774, 1776 and 1778 - were noted in the Heworth register. The nearest Roman Catholic chapels were at Newcastle: unfortunately their registers for this period are in one case not extant and in the other incomplete.

*** A.E. claims that his father introduced improvements in the design of waggons and waggon-way. A 'self-acting incline' was a sloping waggonway, either on the surface or underground, so constructed that the weight of the descending full waggons drew up, on ropes, the ascending empty ones. William Chapman, an engineer and inventor, in 1797 patented a rope-making machine, which was later introduced at the ropery at Gateshead Shore owned by his brother.

Robt Errington waggon and waggonway wright.)* He Rode the first waggon from the Ventor Pit, Low main [seam], he Rode the first waggon from the Discovery pit, south of High Felling, he Rode the first waggon from the John pit near the Sunderland road - and what took plase that day I shall menshon hearafter.**

He had his Coller Bone Broke in the Ann pit by a waggon runing a main [= breaking loose on an incline and running down it out of control]. Six different times he had the ribes broke on the Left side by acdicence. He was very healthy to the age of 76 and during the last 10 years of his life was efected with the Rumatis at times.

He declined going down the pit to waggonway in the year 1799, [persuaded] by a singlar deliverence. Haveing been Imployed in the Sunday night making some new road [underground] and seting the Crain in the Ann pit, near 10 Oclock on the Monday morning [I],

---

* The sentence in brackets was inserted in the margin of the text. The date for the opening of Felling Colliery is usually given as 1779, although a pumping engine was installed in 1776, perhaps for the sinking (Galloway 1898, pp.261,294); and 1777 may have been the date of the initial work on the surface waggonway. High Main and Low Main are the names of upper and lower coal seams in the Newcastle coalfield: up to 1810 Felling Colliery worked the High Main seam.

** Robert Errington worked for the Brandling family, the owners of Felling colliery, for some thirty years, and was present at the opening ceremonies of four of the Felling pits. A.E. later speaks of working in the Venture pit in the early 1790s, and the Discovery and Ann pits must have been opened before 1799, probably long before. Like the others, the Venture worked the High Main seam (19), at least until 1803, when the shaft was sunk to the Low Main seam (NEIME 1885, pp.22-23, correcting Galloway 1898, p.391); but A.E.'s linking of his father with the Low Main seam most probably involves a slip, his father having retired in 1799. At this period, when a new Tyneside pit was opened it was usual to have a public celebration: as the first waggon of coal moved down the waggonway to the river, bands played, guns were fired, and the owners and workmen marched in triumphal procession to a field where the workmen were given a free meal and drink (Sykes 1833, 2:10-11,16-17,55). For the anecdote relating to the opening of the John pit, see episode <49> below.

Anthony, got on the rope to come up the pit, and my Farther and one Samewell Brown was in the Loop [below me].* [At the top], I got safe off the rope and the Banksman [= the workman at the top of the shaft] [was] taken howld of my farther´s hand. [But] as soon as their feet tuched the ground, the hook that had brought them up, it Broke and droped between thear feet. And they both nealed down and returned god thanks for his deliverence. On this, I had to take Charge of the pit woork.**

But to return. He wrought at waggons and waggon way all day, and 3 and 4 nights a week he was down the pit, Erning to get mony to bring up his famely. And he gave us all a Education so as to make us fit for buisness. And his law was to us all to be honest, to be Charatable, to shun bad Company, and to keep the Comandmants in a Christin life. And to love each other was the Charge from our parents. May the Blesing of Blesings be with them Boath now and for Ever in the Name of Jesus Christ our Savour our redeamer.

Anthony Errington born <3>

The younges sun of my farthers femely, I, Anthony Errington, was Born at Felling, in the Cheplry of Huorth, in the parish of Jarrah, County of Durham.***

---

* Small hand cranes were employed underground at road junctions to lift baskets of coal ('corfs') on to the waggons which transported them to the pit shaft. A.E. rode up the shaft grasping the rope; the others had one leg through a loop formed by inserting the hook (on which the corfs were hung when the shaft was handling coal) through the chain immediately above it.

** Robert Errington was 67 when he retired: A.E., aged 21 and apprenticed to the trade for seven years, now took his father's place.

*** A.E.'s birth was recorded in the Heworth Chapel register in 1778.

At 2 years of age I was Efected with the Agoo [? ague]. At $2\frac{1}{2}$ years I was very ill, [having been] efected in the small pock and being blind 9 days. The Only medsine was the white of an Egg and Brandy. The scab Come off my fase by enointing with goose grease and this Brought it off with out any pimples. Soon after this I had the hooping Cough and had medisine from a man riding upon a pye bald gallaway [= horse or pony] who Brought me 2 peney worth of Suger Candy and Ordered Cream and Cours Shuger 3 times a day which soon restored me to health.

Put to skool <4>

During my Childish days, my farther made me a small waggon with 4 wood wheals which I took dilight to trale after me, which was Ingrafting in me my Farthers Brench [of trade]. In a fater day [? after day], I was put to Skool to Mrs Thobren at High Felling. And haveing a small stopage in my speach which made me Lisp, I oft got the Lether strap over me. One morning in the Spring, Mathew Belly and I played [truant], and went to his farther, wheare he was plewing [= ploughing] near Friers Goose wood, and we was burd nesting.* We found a black bird with 4 Eggs and returning $\frac{1}{2}$ an hour after, had the mortification of Seeing a Wazell Eating the Eggs which Escaped from us into its hole. The morning following, mistress asked wheare wee had bean, and not giving a Setisfactory answer, we was sentenced to be hugged. That was, a Stronger boy took Each arm over his Shoulder and leaning forward, breaches being Opened, we got the Cat of 9 tails over us severly. And wee durst go no more a bird nesting.

I Closly atended school to read. And when I got

---

* Friars Goose was then a stretch of open land between Felling and Gateshead.

lorned [= learned] to speling, One Thos Nosbet and severell others was standen to say speling, and I sounded the word ´strainger´, which he had not done, and I got before him. At 12 Oclock, he struck me and made my nose blead. On which a batel Comenced, and I proved Conckerror by thirsting [= thrusting] him into a Boghole and made him all dirt. On going [back] to school and mistress being informed by he first, I was sentenced to be hugged as before, mistress not hearing my report. But upon a nibouring wooman Seeing [= saying] that he gave the furst ofence, she repented and made a small preasent to mee - when it was two late.

Burried alive <5>

The Mistress lorned [= learned = taught] me to wright and shortly after I was put to Wm Yollowly of Low Huerth to [learn] figures in Arithmatick. And haveing bean one month at school, at diner time I went into the Church yard to read the grave stones. And thear being a new digged grave, 4 strong lads took me and put me in the grave, One at my feet and One at my head, when the Other said the buriell sarvice over mee and at the same time with the spade put some Soil upon me. The boys left me their. I was then betwean 2 Coffins that was not fallin in and my finger Ends could reach the grass. I set one foot upon the Coffin and Climbing up, the Earth fell in and Jamed me in the Corner of the grave. I cannot discribe what a smell I felt. I dryed my tears and got some of the Earth by with my hands, and got One leg out and by removeing a little more got out of the grave. On going to skool I informed the Master who fell to them with the Cain, and after he was tired he dismised me to go home, haveing such a bad smell about me, being on a Levell with the Dead. I informed my mother of the Case who instently striped me and weshed me from the Crown to feet. And I got on fresh Close, thease [others] being burnt with fire for fear of the plage. On my return to school, severell persons said to me - Thou has cheated the grave for Once.* The fear of being Borried alive made me start [in my sleep] at night for a long time after, and for severell years after I thought I felt the smell of the Dead Corps. The boys left the School for fear of the Reverend Wm Glovers punishment - I never new their names.**

An experiment <6>

Some time after this, on returning home from school, at the Hollihill pit thear was a Cube and fire for the draft of air, which fire yealded thin scories [? scoria = slag] which being burning in the fire was read hot.*** Being prompted by the 3 other boys, what did wee do but Each one put his Excrement upon the scors and left it frying like pankakes, as we Caled it. A short time after, I was afected with griping and was very ill for 2 days. The Other 3 boys being Efected in like maner, being 2 days from skool. I refer this to men of Physick [to explain] by what Power it acts upon the body after seperation.

Singlar case of bird nesting <7>

One day at diner time I went to Hueth Burn, near the Sunderland road Bridge, and haveing [found] 4 young birds in the nest, was turning over all the stones to cach worms. 2 boys turned up a large stone and it being my turn to Click the worms, to my great suprise theare was 2 hagworms [= slow-worms] 16 inches length

---

* The dialect spoken in parts of County Durham today still employs 'thou' and 'thee'.

** The Revd. William Glover, incumbent of the curacy of Jarrow 1775 - 1808. He officiated at A.E.'s marriage in 1798.

*** The 'cube and fire' was a brazier with a chimney above, arranged over a pit shaft to draw air up from the workings below.

which I nearly touched but did not. And Seting to with thin slate stones we Cut them to inch pieces. Our time for school was past and going their, wee had not gone 20 yards when wee saw 2 Ducks come out of the Burn [= brook] and swallow them up. Wee went to school and I told the Master, who instently sent us to the house of Thos Roe. The house being locked up, wee returned to school. The next day we went [back] and informed the houskeeper who had mised the ducks. She with us went to see for them and near 10 Oclock we found them on the East Bank. And they had been making home and had got from the plase 150 yards and was lying 5 yards off Each other in a putred state, Covered with flies and Creaping raptils. I went for the Clark of the Parish, John Gutree, and hee with many Others came to see them. He Digged two holes and Buried them. The house keeper rewarded us for so much truble, as they had [kept] them on the intent to use when the Master had Company, which might have poisind the who'e.

After this a short time, some Schoole felloys and I was Slinging stones from our garters. I did not leave go in time and the stone Came on my Left Eye and I thout [= thought] it was out. I was Blind of it for 6 weeks. But by weshing with the water of the Bathe well every morning I got my sight [back], which was soon as good as the Other [eye]. (The place [in] Felling Dean, south side of the Sunderland road, under the west bank, is cut out of the rock, and walled with huen stone 6 feet depth, [with] stone steps to go down. The well was filled up being two Cold to bath in - a padick [= frog] put in in the month of June Died.)*

A pistle found <8>

One morning as I was going to School, I found one Bear

---

* The sentences within the brackets were inserted into the text later.

barrel Brass Cock, one Glas file Bottle, with some gun Powder in it, and one Brass tea spoon, wich I took up with great fear thinking it was a pistle belonging to some Rober, as there had bean several people Robed on the way from Newcastle to Shields and Sunderland, the roads being infested at night by foot pads. Upon leting the Master see what I had found, he shoed the same to the Constable of the parish, which returned the Opinion that they had bean used as the means of detaning a poor man 2 days before on his road home [and robbing him] of a few Shillings.

A good doctriss <9>

Returning from school one Evening I was Informed my Old school mistress was very ill and that she wanted to see mee. My Mother and I went to see her the same Evening. She got hold of my hand and said I was to be One of her pall bearers. And none of us more then 12 years of Age, wee bore the pall all the way from High Felling to the Hewerth Church near One Mile.* And she was lamented by all who knew her, being a Native of Irland, [and she] was a good Doctriss, scield in Leting Bleed and Driving out venum out of the feet or hands. I saw One boy take hold of a hagworm [= slow-worm] with fore finger and thomb and Instently he felt pain. I perswaded him to go to the mistress but he did not go until the next day when his whole hand was sweld. The Mistress rubed it and it bursted and a Cup full of poison Came out. It was a few weeks before he got beter.

Lorning to swim <10>

in the time of Hueth School. On the Asencion day, the Mair and Aldermon of Newcastle upon Tine goes from

---

* Isabella, wife of William Thowburn, died 26.4.1787 (Heworth burial register), when A.E. was aged eight and a half.

Newcastle to meeat the Tide at Tinmouth Bar and proceads from thence in barges in full Buty of Posesion of the River Tine up to Newburn.* Mr Yallowly gave [us] play and I with many more went to see the Barges. Off the post towards Sheiels, wee [decided we] would Bathe in the River at a plase Called the Yow Hole. After going in, in a suden I got off the ground and was taken in a whirl Edey of the eb tide. The boys seeing my dander [= rage, panic] and indivering by throwing stones to plosh [= splash] me out, I at the same time [was] keeping myself up with my hands. But I was Like to go down. One recolection struck me. I was Either to do as the Padicks [= frogs] did or go down. I instently struck off and swam to Shore. I was feared of bathing [for] some years after, yet had many a time swim a little to practise.**

Singler occurence of [saving a man at a pit] <11>

Wheare the Felling Colliery was woon, stood two main Ingines, One of 9 feet stroke, 6 feet Cilendar, 16 $\frac{1}{2}$ inches the woorking berell; the Other 7 feet stroke.*** The Botum pump splet, and the Colliery stoped for One week. My father being in the [pit] asisting, I went with his Diner. And their being 2 Pumps [raised up the shaft] on the 2 Crabs [= large

* Tynemouth Bar and Newburn are at the mouth of the Tyne and about 15 miles up-river, respectively. An account of this traditional Ascension Day ceremony can be found in contemporary histories of Newcastle and there is a contemporary painting of the scene (Brand 1789, p.37; Mackenzie 1827,p.744; Middlebrook 1950, after p.148).

** In fact, A.E.'s only other reference to swimming is to an occasion, many years later, when he had to swim in a flooded area underground (43).

*** The 'engines' in this anecdote are pumping engines, used to draw water up a shaft out of the underground workings, the first stage of the pump being underground, at the foot of the shaft. At which pit within Felling Colliery this episode occurred is not stated.

capstans] whith 4 horses drawing, Sopworth Morley, Ingineear, was upon the pump top giding them up the pit.* When near 10 fathum [= 60 feet] of the top, the horses being nearly tired [out], haveing to delve them [= dip their heads with effort] sore [= severely], his brother Christifor Morley had to mind [= look after] the pall [= pawl, a device for locking a capstan], he being driving the horse[s], [lest the horses stopped and the pump's weight took over].** I, standing off at one side, saw the north Crab rope and [noticed that] One strand gave way betwean the ful sheave and main Crab. I instently shouted [to them] to hold the Crab, the Rope was breaking! Christopher Morley made a Jump of great length to put the pall in and stood trembling every bone of him. Mr Thomas Barns, the Viewer, being neare, Came and Inspected the rope.*** He ordered Sopworth Morley to Come off the pump top and Came to Me and Lifted off my hat and Claped my head and said he would make a man of Me. He asked who I was and My farthers name given, he did not forget me, I being [at] the first pay paid hauf a Crown for the word of Comand. Having acted so, I was alway beloved by Sopworth Morley and many a pot of bear have been rewarded with by him.

Saving men on the waggonway <12>

Near to this time, My farther sent me with a Lettor to

---

* The 'crab' was a capstan worked by horses, winding in a rope over a pulley, and employed to draw loads up a shaft. In this instance, the heavy parts of the broken machinery were being wound up the shaft, with an 'engineer' riding them, to guide them and shout instructions.

** The pawl is a wooden or metal bar inserted in a capstan to lock it.

*** The 'viewer' was the manager and head engineer of a colliery. A.E. respectfully refers to all viewers as "Mr". Thomas Barnes was the most famous viewer in the North East coalfield at the very end of the eighteenth century. He worked for the Brandlings for some years and died in 1801.

One Mathew Gray, Waggon wright, at Shirif Hill Colliery.* The waggonway lay near the Windmill Hills and went down the north side of the hills to the Rivir Tine, and at the Coal steath [= staithe] Mathew Gray lived. I was about hauf way down the bank when thur was two Waggons Coming after me Amain [= broken loose and running away]. The later ones Convoy [= brake] had broke. I ran up the hillside and their being some Old timber lying, I throwed some pieces a Cros the rails, which throed the first One [i.e. waggon] off the road. And roning Cros the bye way, the After One struck the Corner of the furst and Intangled, when boath of them stoped in the by way guter [= ditch], 10 yards off 2 light waggons and horses and men, which was in great Jepardy of their lifes. In this I also was rewarded, by Math Gray and the 4 waggonmen, of a Drink of ale and hauf a Crown.

To be a blacksmith <13>

I then being 13 years of Age, My farther would put me to trade. I went One week to the Cuntry shop to bee a Black smith. On the Friday my Master Ordered me to take the Shoes off an Old greasy heald horse, the smell of which made me throw my diner. At 4 a clock the farmer and Master being in the 3 horse shoes, Bublok, I went and told the Master the horse was striped and that I would never be Blacksmith if that was the work.** He gave me a drank of ale and said the throing would go off. But I kept throing some time and at last my Mother asked what was the Cause. She thout [thought] it did not agree with me and I returned to School.

---

* Sheriff Hill colliery was two miles south of Gateshead, and its waggonway ran down to Redheugh, west of Gateshead.

** I cannot identify "Bublok", but since A.E. apparently returned home at the weekend it was presumably not far from Felling. The reference to "Cuntry shop" may indicate that there was some connection between Felling Colliery and this blacksmith's shop.

A pit fires [= explodes] <14>

During this time, at night I had to go down the pit to repair waggons and waggonway and was frequently 2 or 3 nights in a week down. In the Venter pit one Night, My farther and I was going in the way to [the] East, and at the 5th Siding the switch wanted repairing. Their was a trouble [= break in the coal seam] in the mine of 2 feet rise to the East, and some drops of water fell from the top.* My farther was moistening the Clay for the Candle when the gas that had gathered above the Levell of the Air took fire, and went from us to the East.** At this wee Steped into a sump [= drainage pit] up to the Nees in water. Near 2 yards further out then the trouble, the fire returned back to the dyke [= line of seam-break] and went out, Leven us in darkness 500 yards from the pit Botum.*** Wee

---

* The break in the coal seam was such that a two-foot step-up between the west and the east sides of the seam existed along the line of fault, and it allowed the escape from the coal of permeated water and gas. This minor break was probably a ramification of a major fault, the Heworth Dyke, which ran through the Felling Colliery working area (Bailey 1810, p.31), and was no doubt the 'dyke' alluded to by A.E. here and elsewhere.

** The miner's candle was fixed into a ball of clay, both when carried in the hand and when standing by itself. A.E.'s father apparently held his candle up to the roof to catch the drips of water to moisten the clay, but he had not foreseen that gas was also escaping and had gathered in the roof, and the gas caught fire.

*** Methane gas can be produced within a coal seam, and being lighter than air, it concentrates near the roof of a working. When not dispersed by a rapid air flow, but mingled with a limited quantity of air, as can happen if it is concentrated in a confined space such as a coal mine tunnel or a coal face, the gas will take fire when it comes into contact with a naked flame. Although the explosive force of the combustion is limited, a sheet of fire burns the humans it meets; and both the explosion and the fire-flash are considerably augmented and extended in space if the air is laden with coal dust (a point not appreciated in the nineteenth century). Hence A.E. and his father dive into a drainage-pit, below the sheet of fire, which passes along the tunnel above the level of the air. However the explosion had blown out their candles (or perhaps A.E. had quickly extinguished his) and they dared not relight them because of the gas in the air.

then made [along] the way to the shaft. [There was] no [other] person in the pit but One man, Wm Rodgers, keeping the furnis.* He knew what had hapned us and was Coming to see for Our safety and returned [with us]. When [we found that] the fire had Blown down 3 doors [acting as air locks] and some Bretish [brattice = partition to divert air-flow]**, Wee got them put right, and when the Overman [= shift foreman], John Brows, Came down in the Morning, he would not believe us that any such thing Could take plase. We was not down the Next night and he, being the first that went in the same plase, it fired, and he was sore burned, face and hands and breast, and said, "Who whould have thout [= thought] that!"

To be a farmer <15>

After the pit firing, I did not like going down the pit. [So] My Farther put me to be a farmer. I entered the week before Christmas to be a farmer Near Hilton Castle, County of Durham, and was their One week.*** I did not like my place. I had to lead turnups upon a sledge with an Old reasty [= rancid]

---

* Coalmines had coal-getting shifts and maintenance shifts, the latter usually at nights or weekends. Thus we find A.E., a waggonway wright who mainly worked underground on the maintenance shifts, working at nights; and the pit on this occasion empty of other workers.

** The ventilation of the pit turned upon a single current of air being conducted or 'coursed' along the various tunnels and faces, the current being directed by a series of doors and partitions so that it swept round the whole workings before returning to the surface. After an explosion, the collapse of air-doors allowed the air to short-circuit, leaving parts of the pit without adequate ventilation and hence permitting gas to concentrate in them. The sluggish natural circulation of air through a mine was encouraged by passing the returning air over a furnace at either the bottom or the top of one shaft: the heated air rose rapidly up the shaft or up a chimney, thus drawing in fresh air down the other shaft.

*** Hilton Castle is about three miles west of Sunderland and about eight miles from Felling. This episode probably occurred in the winter of 1791-2.

horse up to the knees in snow. Being only upon triell, I gave farming in.

Condemned to walk the plank <16>

I got aquanted with the Boys of the Felling shore and One Wm Hall who worked at the Steath [staithe = waterside coal-loading point] and I was Invited to go in a boat on the Setorday afternoon to Sheilds. Which I did, and got some Mushells at the Lowlights and Came into the harber.* [Then] the tide flowing, we made [our] way home. When we got to Jarrow quay East end, their was a fight took place in the boat with [= between] Wm Hall and the Boy the boat belongd to. 11 boys against us two drove us to the head of the boat. A board being plased Over the side, wee two was Condemned to walk the plank. On jumping in, we was up to the shoulders in warter on the Edge of a large Sand Bank, and taken no notice of the Tide flowing, we had to swim from the sand with Our Close on to Jarrow quay [where there were] 2 Ships lying by the quay. The Boys in the ships, seeing our treatment, felt for us and gave us Beef and Biskets to Eat and fill our pockets to set us home, and we reached home at 9 oClock the same Evening. And repented going a pleasuring.

Apprentice to my father <17>

At 14 years of Age I then formed the resolution of being my Farthers Brench of Calling, Waggon and Waggonway Wright. And I started to Clean the way for 8d per day, and then again going with my farther and brother John, in a short time the fear of the pit left me. And with being in varius parts of the mine, I understood the Carrying of Air [= coursing of the air

---

* The boys go down-river to the Low Lights at North Shields, near the river mouth.

through the mine] - yet never had anything to do with that part. I was taut the pit Language and got on with my trade very well.

I had worked 2 months with my Farther, and [had been] working with the Axe and adge [= adze] and augers, [when] their was a Ship Builder at Felling Shore in the name of Doegg two or three times [was] perswading me that to be a Ship Wright was a beter trade.* I at last named [= mentioned] it to my farther, and on this I agreed to be bound Aprentice to My farther. Their was no time lost and Thomas Robison, Taylor on Gateshead Fell, Senier and Junier, was Bondman. A fue days After, the Ship Builder Came to me and would wish to have me. I told him he might leave off for I was determined to be my farther´s trade.

Working with the adge [= adze] <18>

Mr Chepmans patent laid rope at the South Shore rope walk near Gateshead was to Comence making. My farther had laid a wood railroad for the Waggon to travell upon, [with] men turning handles upon the waggon, working touth and pinion wheals, [which] Laid the first day 20 fathoms of Ships haus [= hawser] rope. The whole length of the rail road not being finished, on the Setorday afternoon I went to Asist finishing the way and working with the Adge, right hand and Left hand, I was taken notice of by Mr Chepman and Others, and for my Activity was presented with 3s 6d and to have my super and Lowens [= allowances]. The whole of [the] Korvers and Machenicks on the Tine and Wear was preasent, and when One fathum was laid, there was Much regoising and hesazes [huzzas] and guns firing. There was 21 Barrels of 8d Bear given to the Peepel, and all

---

* "Alexander Doeg, shipbuilder" (Directory 1811, p.53), a Scotsman, hence, as a Presbyterian, eventually buried at the Ballast Hills burial-ground (Mackenzie 1827, p.412).

men rejoised at the Improvement.*

Singlar profisy <19>

During the five weeks my farther was at South Shore, I Anthony had to go down the Venter Pit to repair way and waggons, and no one with me. William Rodgers was furnesman and he Came in to bear Company. For $\frac{1}{2}$ one hour he took me off the railway into the waste and Said to me, "Anthony, wee will go to prayer". He prayed to God through our Lord Jesus Christ and [seeing] that I was silent after prayer he gave me the Book to Kiss with [a promise of] secrecy untill that should be fullfiled which was shoen to him in a Vishon, which he had to revial to none but me.** That [vision] was, that when the Berriers [of coal] was worked off in the Later part of the High Main [seam] that wee was in, the seam below would be Sunk to, and their would be Hevey Misfortunes. And that I would Escape.*** "Some of [the] Mine wall escape and some fall here." I had to kiss the Book again [and promise] never to speak on that until it was fullfiled, which [promise] was kept.**** We then

---

* There is some confusion in the chronology at this point. The present episode is depicted as occurring shortly after A.E. was apprenticed in 1792, but the work at Chapman's ropery was in fact not undertaken until 1799, as stated earlier <2>. "Korvers" made and repaired 'corfs', basket-work coal containers used underground.

** The "Book" was presumably the Bible, and William Rogers seems to have been a protestant visionary - his name is not in the Newcastle Catholic registers.

*** Barriers of coal were left between sets of workings of a coal seam, in order to support the roof and also to hold back water or gas. When the working of a seam had reached its limits outwards from the shaft, some barriers were reduced or completely removed, in a movement retreating towards the shaft.

**** Writing in the 1820s, A.E. intended this prophesy, supposedly made at an unstated date in the 1790s, to refer to the famous 1812 disaster at Felling, when 92 miners were killed in an explosion. The Low Main seam in which the explosion occurred was reached at Felling in 1803 but worked for coal only in 1810. A.E. survived because he left Felling in 1811.

returned to our duty and was allways good friends.

The powre of Christ reineth <20>

One morning [at] 6 oclock as My Farther and I was just gone out of the [house], there was an Aged women coming with 2 jugs, One for Old Milk and the Other for buter Milk. My Mother and 2 of My Sisters [had] started at 4 Oclock in the Morning, and they was [still] Chirning. This Old woman set her jugs down And Came past My Farther and I, and Said, "Thee mite chirn away there...".* I went home to breakfast and took My Fauthers down [to where we were] working on the railway. At 9 Oclock My Sister Jane Came to My Farther [and said], "You have to set Brother Anthony home directly. Mother Cannot get no buter". [Father said], "Go Lad, and God be with thee." I run home and was Out of Breath. Mother in tears Called me into a small room and told Me what to do. By this time there was 10 Or 12 women wating for Milk and Buter. I went up to the Barrel Chirn and took hold of the Handell and Said, "Depart from Mee, O all ye that work Inequety, and Let the Poore [= power] of Our Lord Jesus Christ rean [= reign], in the name of the Fauther, Sun and Holy Goast". And turning 3 times from Me, first the second turn I felt the Buter on the Breaker, and turning 3 times back, the Buter was flap flap flap on the breakers. I stoped and Mother was in tears. I said, "Mother, heare is the Butter". She came and said a short prayer, which was, "I thank thee, O God, thee hast Menifest thy Poor [= power] above the Enemy". The Nibouring women were astonished. The Event was Every Ones talk.

The seccond case <21>

One Month after this, Henery Stobbs, Farmer at

---

* The old woman, piqued at not getting her butter milk, put a spell on the butter making.

Carhill, Came down to Felling Shore and asked my fauther for Me to go with him, seing [it was] near 10 Oclock, [yet] his Mother could not get Butter. My fauther said, "Go and God be with thee". I got on to the horse behind him and was soon at Carhill. The kitchen [was] full of women and men. I went to the Chirn which was Much bigger then My Mothers, I drow the Cork and Breathed into [it]. [Then] I put the Cork in and took hold of the Handle. All was Silint. I said, "Depart from Mee, O all ye that work Iniquety, and Let the Poor [= power] of Our Lord Jesus Christ reighn, in the name of the Fauther, of the Sun and of the Holy Goast". [I made] the Sighn of the Cross and turning 3 times from me, I felt the buter, and turning 3 times back again, the Buter was flap flap flap. I gethered it a Little more and the Mother took out the Buter. She ofered me some Money which could not be taken [by me]. The Man and Women said it was a Miricle. I took from her hand a peace [of] cake and a pot of the Buter milk, and returned home.

Third case <22>

6 weeks after this, I and Fauther was at Felling Shore working when [a] Man and Horse Came for Me to go to Squire Rusells of Low Heweth.* The Houskeeper [there] Could not get Buter, haveing Chirned five hours. [It was] near 10 Oclock when I arived at the Kitchen. The Squire did not go to Newcastle at his usual time, 9 Oclock, and he was preasent. I asket the women, had she put any warm Warter in it, and she said, "Near a gill". I drow the cork and Breathed in to [the barrel]. I then took Hold of the handel. All was Silent. And I Said, "Depart from Me, O all you that Work Inequity and Let the Poor [= power] of Our Lord Jesus Christ reighn, in the Name of the Fauther, Sun and Holy Goast". I turned 3 times from Me and the

---

* The Russells lived at Heworth Hall.

second turn felt the buter, and turning 3 times back, the buter was herd by all preasent to flap flap flap on the Brickers [= breakers]. Squire Rusell was Standing 2 yards off with his hands to ward heaven, and he said, "Good God, what A Miricle!". "Come in hear", the Squire said, "to the next room", and asked Me, would I have a Little Cheas and Bread and a drink of bear. Which I took in the Squires Company. He ofered [me] some Silver but I told him I Could not take money for Mentaining the Poor [= power] of God. He shook hand with me and said, God be with me all my days.*

Brotherly love <23>

Brother Geord[i]e [en]Listed to the 15 ridgment [of] foot. And had bean in that 3 years [when] on the Sunday Morning I went to Sunderland Barricks to See him. When I got there all was in a bustle. On Setarday the route had Comed for the regment to March for Irland on the Monday Morning. I made a preposition to the Sargan for My Brother to [come] and stop all night at fauthers house before he went away, and I would Carrie his gun [home]. The Sargant said [that] I would go [with him] to the Captan. Wee met the Captan and told him the request. [He said] we was

---

* Persuading butter to set was a common duty of English rural magicians. The following magic rhyme was still used in some districts in this century: "Come, butter, come, / Come, butter, come./ Peter stands at the gate / Waiting for a buttered cake / Come, butter, come". (See the chapter on 'The witch in the dairy', Kittredge 1928.) This rhyme illustrates the connection in the popular mind, in the post-Reformation period, between the Old Religion ("Peter at the gate") and the Old Magic. As Bishop Corbett put it - "By which we note the Fairies / Were of the old Profession". No doubt the relatively few Roman Catholics in eighteenth-century rural England were regarded with some awe by the predominantly protestant peasantry because of their mysterious gestures (breast-crossing and genuflexion) and wonder-working objects (holy water, relics and medals). Hence A.E. becomes a magician: all he does is to make an un-protestant gesture, "the Sighn of the Cross", over the butter barrel.

all to go with him to the Cornell. The Captan related to the Cornell my request [and the Colonel said], "I see, the scheame is for your brother to Desert". I answered, "And he durst not be a deserter, he **would** be disowned by fauther and Mother, Brother[s] and Sisters". On this he [consented and] the Captan wrote a pas for One Night, [my brother] to [re]joine the Redgment at Felling where they had to pas the house [on their march]. On this, his bagage [tied] al up, we started at One a Clock PM and reched Felling at three. The Whole Femely and the Vilage was uplifted at the respect given [us]. Fauther Laid the plan for the Ridgment Coming in the Morning at Nine Oclock, 1500 men and oficrs. Brother was in the Last Company, or Light Company, and when he Stept to his place, Fauther and Brother John were on the right hand of George and I was on the Left. Fauther wished the Captan to halt, and All the Oficers was to come in to the Cotage to take a refreshment, Cheas and Bread for the Comerades and A Cold Table for the Oficers. The Cornell said we was a fine Femely and put down in his Book a Memorandum what my Fauther said [about George], he was to be a Vallient solder, to Concer or Die, Never to Desert. I requested the Captan to have him One night more at home which he granted by taking a Corprell [to stay] with him, which was done. We rose at 5 Oclock, Brickfast on the table. I Accompained them to Newcastle [where] My Brother and the Corporell was plased Reargard, and I set them to the west gate. Brother said, "Heare, theire is a Shilling, go with us!" The Sergent Secconded it by the glasses again, which I paid for. But I would not take the Shilling. [I said], "Should My Brother die in battle, wright for me and I will take his gun". He wrote this down and bad fare weel. [But] the Sergent run and told the Captan of the gard who called [out], "Halt!" - A. Errington had volenteared to the Redgment! On this there was 3 Chears, which made the Horse Ridgment at the Barricks to Sound to Arms and which came full gallop thinking the Ridgment had Mutienied. Which was a mistake.*

It was 19 years before we Could recieve one Leter from him. 24 years after [he left], he came home, not One [w]Ound on him, nor Lash on his back. Out of 1500 men at Newcastle when he left, there was 26 Left [? alive] behind [= ? besides] him of that number. They was a short time in Irland, and [then] went and took the Island of Barbadus.** 300 died [there] of the sweting sickness. He had it for 6 weeks, and had 3 pints of rum the first week per day, after that had 6 pints per day, sweting night and day. Thease that could not swet [it out] died in a fue days. He was discharged [with a] Sergents pension of £25 19s 10½d per year. Whereon I was not Called upon to serve. He lived 13 years after and was buorried at South Shealds.***

A stolen gallaway [= horse or pony] <24>

There was two Brothers, wevers by trade, and Dwelled in Yorkshire. At Braugh Hill fair, the [Felling colliery] hors keeper, William Turnbull, was there and

---

* It seems likely that in this episode, as in certain others, A.E. was having his leg pulled and did not realise it. Perhaps he was regarded as a rather simple fellow. The story of the horse regiment sounding to arms because they thought the infantry had mutinied sounds like an invention to embellish the anecdote, yet it is not out of keeping with the military history of the period. This episode almost certainly occurred in 1793: in 1795 several foot regiments stationed in Newcastle mutinied and paraded in disorderly fashion through the streets, the men chanting their grievances to the townspeople (Sykes 1833, 1:376).

** A.E. was wrong here. Barbados, in the West Indies, was already a British possession. But it was the military base for the successful expedition of February 1794 against the French island of Martinique. The 15th regiment of foot had sailed from Cork to Barbados and from thence to Martinique (Fortescue 1900, 4/1:351,354).

*** Since George died in 1834 (South Shields burial register), according to A.E.'s figures he retired from the army in 1821 and the regiment left Newcastle in 1797 - whereas in fact the regiment left in 1793.

Bought 6 gallawas and Brought these to the Colliery and put [them] down the pit. One of thease was the Brothers gallaway. They got Intiligence of the gallaway being sold at the fair and [that it] was gone North. [So] the two Sets off and got to the Black Bull in Gateshead, and the Landlord directed them to William Turnbull. And they Came to the Venter Pit. And Turnbull requested me to go down the pit with one of them, he with the other. These men had not bean down a pit before and was very feared. The Stabels was 60 yards from the [pit] Botum, and the large furness [they passed] Reminded them of Hell. One of them said [of the gallaway], "If it dose not Neigh when it hears my tonge, it is not Ours". We went forward and on Entering the Stable, [where] there was 16 [stalls] and it was [in] the 17 Stall, it Nied direct it herd him. And when the brother aproached, it Licked him with its toung. He got his arms round its neck and they boath Shed tears of Joy, and Said, How in the world [would] they get hit out [of the pit] ? I and One of the Brothers asended and put the horse net down, and got it safe up [to the surface]. They rewarded us with Chease and Bread and Ale and they returned to their home with the Gallaway.*

The tayler´s goose <25>

I, Anthony, had to go to Newcastle upon Tine On Erend for My Farther. When I got to the North End of the Bridge, Thos Robison, Ceinier [Senior], Informed me that there was a Wager of 20 guinies that a Strong man, a taylor, had to throw a taylors goose [= flat-iron] from the Hauf Moon Battery over the houses and

---

* This 'galloway' was apparently one stolen from its owners, either before or at the fair (the punctuation of the relevant clauses is uncertain). It was purchased in order to be employed underground at Felling Colliery. Throughout the coalfield, horses and ponies were extensively used underground (as ponies continued to be until the 1960s). They were stabled underground, being conveyed down, and brought back to the surface again (usually only when dead), in a horse-net.

Shops to the Bridge End, Leave [having been] given from the Mayr of Newcastle, and Constables sent to Stop the streat.* Word being given, the Man flung the goose away. But on his delivering the goose a man juseled him and the goose fell in the north side of the ridge of the house. Then said Thos Robison, [I should go for it and] he would be Bond for my honesty.** On this, I was to go with the Shop Boy to the garret [of the house]. 2 Slates [were] Broke near the chimney and wee had Several Boxes to move before wee could [find] it. On going to the door, the Bridg and Streat [was] full of Men. The cry was, "Heare she is! Here she is!" Being taken to the Battery and [handed over, the goose] was Once More [to be] flung by the same Man, the road being Cleared and Streat Stoped. The call was, "Let flee [= fly]!", when she Cleared the ridge, and Easing off the house [by] three feet, [fell] and Let [= alighted] on the Pavement and Flew 5 yards [along it]. [Thus] the second flee the Taylor woon the 20 guineas. I had to go with Mr Robison to the 3 Buls Heads to have a drink of Ale. And he said I was a Luckey Lad, and hily respected by the taylors.

Singlar meeting <26>

One Sunday Morning, I was going to Church in Newcastle and going along the Sandhill, I saw a man standing on the flags [= flagstones], the Very Picture of my Self, being in One iniform [uniform] of Dress, Complection, and hair allike. He Looked and I Looked, and [I] said, "I think I have found my Shaddow". The bells of

---

* Newcastle Bridge was then the only bridge across the lower Tyne. Half Moon Battery was a locality near the bridge, formerly the site of a gun emplacement and later a tenemented building (Mackenzie 1827, p.168). In 1788, for a wager, a man threw a stone over the 194-foot steeple of St Nicholas' church (Sykes 1833, 1:348).

** Thomas Robi(n)son was probably the same man who was bondsman for A.E. when he was apprenticed <**17**>.

St Nicklos Started the first Peal [as] we Entered intc Conversation and walked along the Close, [then] past the Infirmery, and we got to the Read Lyon at 12 Oclock Opisite wheare my friend dwelled.* The Farther and Mother and a younger Brother was at Church, the sister was Cook. He Laid the plan of Interview. As soon as they went along the pasage, I was to go in and [through] the first door to the right hand, and turn to the right, [then] set my hat on the side table and [I would find] Diner on the plate. I [did this and] sat down and got 2 bites when the Mother said, "Wheare was thou, Tom, I did not see thee". Being Eating I did not Answer, when [the] Sister said, "Mother, that is not Our Tom!" The Mother and Farther said, who [was it then], she was wrong, [but] upon this the real Tom Entered. They boath dropd Knife and Fork. Tom said, "I have sent my Shaddow to Diner". Tom sat down near me and it was hard to tell which of us was ther own [child]. This Farther, Wm Bilton, was a Mill wright, and boath suns Millwrights. We spent the day

---

* A.E. lived in Felling but the nearest Roman Catholic chapels were in Newcastle. St Nicholas' church, not far from Newcastle Bridge, was the (Anglican) mother parish church of Newcastle. If its services were still at the same times of day in the early 1790s as they had been in the 1720s (Mackenzie 1825, p.291), then morning service began at 10 o'clock, and hence its bells would have commenced to peal about 9.40. If this was so, then A.E. was indicating that he and his new friend chatted for some two hours while walking to Tom's home, a distance of little more than one mile, even by the round-about route they took. From the Sandhill, a "spacious, well-paved area" near the end of Newcastle Bridge, the boys proceeded westward along 'The Close' (still so-called), parallel and near to the river. They then turned north, to pass the Infirmary on the Forth Banks, and through fields reached the West Gate (Beilby's 1788 plan of Newcastle; Mackenzie 1827, pp.161,165,196,291). Within the town again, they went towards Newgate Street, where the 'Red Lion' was located (a less likely 'Red Lion' was in St John Street) (Directory 1787). In the 1780s, one Roman Catholic chapel was situated in a court off Newgate Street and another at the foot of Westgate Street (CRS 1936, pp.203,206). Having been diverted from attending Mass this Sunday, A.E. took a round-about route which was perhaps chosen to avoid passing near either chapel. Tom Bilton, apparently also in his Sunday best, and encountered by A.E. when also some distance away from home, was perhaps heading towards St Nicholas' church; and if so, he too was diverted.

together and agreed that he was to go to My Fauthers the same way. I instructed him and he went and sat down and [had] begun to Eat when My Sister Jane said, "Mother, that is not our Anty!" Fauther and Mother said, "Lass, thou art Out of thy mind". [But] she insisted on that, upon which I entered and said [that] I had sent my Shaddow before me. We got Diner Over and Mother set the botel on the table, and we Ingoyed the afternoon in socichel friendship. I saw Tom twise after. I shall relate heareafter what Occured.*

Manchester waggon and eight horses and drivr and poney <27>

I, Anthony, upon the Setorday [at] 2 Oclock PM, was going to Newcastle to meet Fauther and Mother, on the Newcastle fairday [in] August.** There was Pasing the Sunderland road [in] Gateshead, the [Manchester] Waggon heveyly laden from the South. I walked Sharply to overtake the Waggon, and near Jackson Streat I came up to it. And the Wheals being nearly Eyteen inches bridth, I saw the Far hind wheal wreggel. And stooping and looking, [I saw] the Axell was broke and had [moved and] was five inches worked out. I instently Cald, "Stop the waggon! Your arm is Broke!" The Driver hearing, said to the horses ["Whoa!"]. They stoped, and instently he came wriding to me and off [= from] the Poney Looked [at the axle] and said, "Good God! What Providence!" Next he said, "Whats to be done?" I answered and Said, heare was a Smith Shop Close at hand. I went for Thos Hedly, Smith, [but] he was at Newcastle and [only] a boy in the Shop.*** I said to the Waggoner that I was a waggon wright and I would Asist him. I then went down streat to a

---

* There is nothing further about Tom Bilton in the autobiography as we have it.

** This would be the Lammas Fair, on the first Saturday in August.

*** "Thomas Hedley, smith and farrier, Gateshead" (Directory 1811).

Publick house where 9 of Mr Haukses Smiths was, [having] done their weeks woork.* I requested 6 of them to go to Hedlys Shop and Mend the Axell, promising pay and Lowens [= allowances]. The Men Jumped up and Set to and made a good Job of it, being [made] Stronger then it had bean. The Waggoner paid the men and [gave] Lowens too, and then Ofired pay to me. I obgected and said I was a Prentice and could not take Money nor pay for Gods Providence. On this, he was Astonished. We got all Clear from the Streat and the Waggon proceded on to Newcastle, being detained [only] 2 hours. 4 months after, My Fauther and I was in Gateshead and the Waggoner and his Master was there. I went to see how the Arm was. He said, "Quite strong. Stop, my Master is heare", and they Came cros the streat to My Fauther. The Master shook hand with My Fauther and Said, "You have a wonderful Sun". The Waggon had [had] One thousand pound[s] worth of gloas [? glass] in the Waggon for Newcastle. He would treat us boath, and Insisted of My Fauther to take 3 hauf Crowns, which was done to buie Me a new hat.

Singlar escape <28>

[One day] I was assending the An pit at 3 Oclock PM, One boy in the Loop with me and 8 more was on the rope, above One and the Other. When we were 20 fathums [= 120 feet] from the top the Crank of the [winding] Ingine Broke, and we fell back 2 fathom. I instantly grasped the boy with me and Said [to the others], "Clag to [= stick tight, i.e., hang on], boys!" They did so and all kept hold. One quarter of the Crank Eye fell down the pit, Missing [us] all. [It weighed] 2 hd 0 qr [= two hundredweight], and Carried away the Scafold at the botum. On this, I Called up that all was safe, [and that they should]

* Messrs. William Hawks and Co. owned an ironworks at Gateshead, hence presentday Hawks Road.

run down the [rope of the] Crab [= the horse-winch]. The banks Man [= workman at the pit top], William Anderson, Called for Help [from] Al hands [there] and 20 men run down the Crab [rope]. [I told] the boys [that they] was not to touch the rope unto it Came to me. I Stedeyed the [crab] Rope and the boys Changed Rope One after the Other. Thease [being] drawn up, the Crabrope Came [down again] for [the] Boy and I, and wee got Safe up.* But being standing speaking with Anderson, he said, "Stop, thou are Lamed!" The Blood was running down my Side over the top of my trusers. I striped and found I had bean Cut on the wright Shoulder blade to the bone which bled freely. A women ran and Brought a Cup ful of Brandy and Loaf Sugger [which] stoped the Bleading. This [w]ound healed with out any more [trouble] but Continued for 20 years a read Lump the sise of a hors bean. We returned home, thanking Our God for deliverance. I supose that the Cut had bean done by One of the Iron Wedges.**

---

* The cage in which men later rode in the shaft had not yet been invented. The miners descended and ascended scores or even hundreds of feet either by grasping the winding rope with hands and legs, or by straddling the hook or a loop at the end, as A.E. and one boy here do. The risk to life in this episode was considerable, since there was a danger of falling a great height, not only as a result of being jerked off the rope when it suddenly fell back twelve feet, but also as a result of losing hold when transferring from the first rope to a second one dangling in the shaft. The depth worked by Felling Colliery was said in 1787 to be 70 fathoms (Galloway 1898, p.295): if the Ann pit was this depth, then A.E. and the boys were suspended 300 feet above the pit bottom.

** Here, as at a few other points in his text, A.E. inserts a passage which provides a detailed account, in technical language, of certain aspects of colliery engineering practice in his day. The present passage, being a digression, is relegated to the footnotes and appears below. It seems likely that A.E. kept a work diary and that he copied technical details from the diary into his autobiography. Here he details why the Felling winding-engine broke down. An underground drainage system failed, with the result that the surface engine, which worked pumps as well as doing the winding, raced in its pumping capacity and broke a crank. Some of the terminology is archaic which makes it less than certain that the editorial insertion of punctuation has been done correctly.

"N.B. The Ingine Landed One Corf [= basket of coal] per Minuet,
<continued on the next page>

I, Anthony Errington, John Hall, Mathew Sanderson and Thomas Hutchison, we had to go from Felling to Gosforth Colliery, a land sale [colliery], to repair the railway, and on the Monday the Coals was Sold near the Moor Edge, Morpeth road.* John Hunter was agent and had a Brother in the Old Flesh Market [who] kept the sighn of Dog and Duck. [? Hunter said] wee was to go there and have One gallon of Ale. There was 10 men in the house, tradesmen, and one [of them] was Singing one of Burns Songs when their was a Man pased me and sat down on the End of the Bench. Shortly after I saw him Sobing in tears. I went to him and [said], "Whie are you so sorrowfull? Wee are all cheorful". He Answered he was the Author of the song. I instently went to the table and Said, "Robert Burn in Company, Jentlemen!" One [man] Said, "I now [= know] him", and on this he came and Shook hands with him. He was ushered up Stairs and Mary Morchents sent for the Barber to shave [him]. And he was Striped and Clad

---

<continued from previous page>
3 score per hour, 12 hours per day, and this 33 score was the days work, [allowing] One hour Changing men. The pit had bean at work Every day Except Sunday for 3 weeks. There was 2 sets of pumps Lifted the warter from a drift [= tunnel] 20 fathom below and delivered [it] to 2 Large Ingines. There was a Staple [= underground shaft] 100 yards from the Coal to the Drift, at the botum of which was a Dam and a horse foot Valve, 10 inches Diametr. This valve was worked by a wood Spear and Leaver to the Stone round the top of the Staple. Not being secured by good props the Stone gave way, taking fraim, Leavr and Spears down to the botum, and the valve being [en]gaged Open Only admited part of the feeder [= water flow]. And when the Ingine worked reglar, the Ingine strok[ed] on air. And not having the weight of the Colum of Water - the flywheal was 27 feet - the velosety obtained broke the Crank. Some Ignorent Men Said the Brakemen Could breack the Crank when he Wished. Seldom One Lasted more than 3 weeks."

* Gosforth is immediately to the north of Newcastle. A 'landsale' colliery sold all its coal locally and not by shipping it 'seasale'. Gosforth Colliery therefore had no waggonway to the river but it did have one to the nearest road. The colliery was owned by the Brandlings, the owners of Felling, A.E.'s employers, who later connected it to an existing waggonway and converted it to seasale. In 1817-1818 A.E. was employed again at Gosforth.

all in New in $\frac{1}{2}$ an hour. Super [was] Ordered for 16 men [at a] Subscription of 2s 0d Each and wee was not to Leave the Company. After Super, Burns was requested to Sing his favoret [song], Sweat and Lovely Jean. At 3 Oclock in the Morning [this] was so Efecting to the hearors there was not One drie Cheak in the room. Each drunk what he Liked and the Cost was 6s 0d to each, super and drink. At 4 Oclock wee Broke up and returned to our deuty at Gosforth.*

Detecting a theaf before sunrise in the morning <30>

Samiewell Haggerston was night Watch at the Discovery pit, above High Felling, and hee came down to get Super at 11 oclock. There was a Stile by the railway gate. And One Mary Hall, Wesherwoman, saw him pas home, [and she knew that] he returned to the pit the Other way, [so] she thought she would get One Swill [= tub] of Coals in his absence, and did so. She came with the Coals and the Swill upon her head to the Stile [on the path, but] she Could go no further, and Clear of her Load Could not get. And ther she had to Stand to 3 Oclock on the Whit Sunday Morning. Brother John and I was going up the railway and we saw the women Standing and [the] Coals upon her head. Haggerston [saw us approaching the stile and] was Coming down running and Shouting, "Stop!" [He had] to go down to work that morning, not to go at night [as he normally did]. He Came and took a paper out from under the Stile, [whereupon] the woman tosed the Coals

---

* The Scottish bard would certainly not have been out of place in such an alehouse scene, and this was known at the time: "the late Burns was addicted to the bottle" (Newcastle Chronicle, 13.6.1801). However, this must have been an impostor. Burns' only visit to Newcastle was in 1787 (Roy 1985), while this episode must have occurred in the later 1790s. It was probably 1795 or 1796, since Burns died in the latter year, and it is unlikely that literate workmen on the Tyne would have been unaware of his death. The bard addressed many songs to 'Jean', but a song entitled 'Sweet and Lovely Jean' has not been traced - 'lovely' was not an adjective commonly employed by Burns (Reid 1889; Kinsley 1968). By the 1820s a Burns Club existed in Newcastle (Mackenzie 1827, p.598).

down, took the Swill with her and went home. Brother and I was within 5 yards when this took plase. He [i.e. Haggerston] told me after that it was us he wished to Stop, the woman was fast eneuth [= enough]. He was scield in Astronemy and understood the Poor [= power] of the planets. This Brother and I was witness too.

To change of life <31>

Mrs Hall kept the Felling gate.* She had only one Doughter, and Charles Purves, Clark at Felling Colliery, was [her] sooter. They came from Alandale. Purves and I got Equented and I was [= ? used] to goo with him to the gate in the Evenings. At this time, Ann Hindmarsh was houskeeper to Mr Sill, Squire Ellisons Stewart.** I got aquented with her and in proses of time I Maried her and we set up house at High Felling where My first born Sun was born.*** 12

---

* This was the turnpike gate known as Kirton's Gate, which was on the Sunderland road, on the boundary between Gateshead parish and Heworth chapelry, just on the Gateshead side.

** The Ellisons lived at Gateshead Park House, one mile from Felling. Thomas Sill, "Land Steward of Kirton's Gate", who died in 1801 aged 87, had been land agent to the Ellisons for over half a century (Heworth burial register; Hughes 1952, p.74)

*** The connection between the Felling gate and Ann Hindmarsh is not made clear in the text. Perhaps Ann was a friend of the gatekeeper's daughter, who was being courted by A.E.'s friend. Or, if she did not live in at Mr Sill's house, perhaps she lodged with Mrs Hall - or else merely passed through the gate each evening on her way home from Mr Sill's. Or perhaps it was simply that A.E. was encouraged by his friend's example to "get equented" with a female. The Heworth Chapel register records the marriage in 1798 of A.E. and Ann Hindmarsh "of the Parish of Gateshead". It also notes the birth of a son, Robert, and A.E.'s wife is here described as "Ann Hindmarsh of Kirkwhelpington", an agricultural village ten miles NW of Newcastle. The Kirkwhelpington register records her baptism in 1768. Thus, at marriage A.E. was aged nineteen but his wife was thirty - and she was most probably not a Roman Catholic, as he was. The autobiography refers to the birth of a daughter in 1800, and four children were alive in 1809. But Robert was the only child whose birth was noted in the Heworth Chapel register, and only one child, Isabella, born in 1806, was recorded in the

<continued on the next page>

monthes after, we Shifted [house] to Low Felling.. The Cause was one for us boath. I had bean down Hollihill pit at Night and I got home at 2 Oclock AM. An Opened the door for me and [as] the fire was dull I Lit the Candle to get Super. Ann related to Me that she had seen a Spiret Dresed in Blue Silk. I told her not to menchon [it] to the nibours. After washing I went to Lock the door. On turning round, I saw the figure of a tall slender Women dressed in scie Blue Silk which walked into the Corner and Disapeared. I kept this to my Self thinking Ann would not dare to Stop at night when I was at work. There was One Emty house at Low Felling which I aplied for, and we got shifted the next day. This poor [= power] of seeing a Spiret, Man or Women, that is boorn in the twie light before the sunrise in the Morning, this was the Case with me and Ann also. This from the Reverend William Warlow, Cathlick Preast, Newcastle on Tyne.*

Hollihill pit blower <32>

The East board [bord = coal-getting gallery] Comen[c]ed and was Droven night and day, and when [it was] 200 yds down, it got a bloer [= blow-out of gas]. Their was a 10 inch brick wall Cloose plastered as Bretish [= brattice = airtight partition] and when the Sound of this blower, which had water along with the gass, [was heard] at 200 yards, [it] was as a Boy beating a drum. It Continued and so [was] called the Drumer. 2 Depeties [deputies = district foremen] and 12 men and boys was [working there], 6 of thease

---

<continued from the previous page>
baptism register of St Andrew's Roman Catholic chapel at Newcastle. Ann was five months pregnant at marriage, which may explain A.E.'s laconic reference to "in proses of time".

* This was the Revd. William Warrilow, who served the ex-Jesuit chapel in Newcastle between 1773 and 1807 (CRS 1936, p.206). Part of A.E.'s title for episode <**30**> (immediately previous in this edition, but not in the manuscript), 'Sunrise in the Morning', seems as applicable to the present anecdote.

men was boaring. I was Alone, repairing the horse road Called pening near 50 yards out by [= out-bye = towards the shaft] side of the Drumer.* [Suddenly] the Drumer had Ceased Beating. I had a standing Candle, I saw it safe, and with the Other Candle in My hand [approached The Drummer]. When 5 or 6 yards off, I stoped and Could Decern [a cloud] as mist Coming down out of the hole where the warter had Come. This was pure gas.** I [acted] with presence of mind to run - to run to the high way to give the Alarm. When I got theare the Depety said it [i.e. the gas] was just Coming and he Called to all men and boys to put out their Lows [= candles]. They and I retreated to the waggonway, [where] we had the scale [= leakage of air] of 2 doors, fresh Air. We Ordered all out to the pit Shaft. The 2 Depeties, Nickell Urwin, Robert Stove, and I went down to see how it was going on.*** It was Silent and had Backed the Air from top to botum 5 or 6 yards.**** There was no time to be Lost, and we run out to the pit botum. The men [there] had Called up [the shaft] but there was no person wating at the pit [head] at this time. [So] I Called up with all my Strength of Voice. William Anderson, the Banksman [= the workman in charge at the top], herd Me into his Bed and said to [his] Wife, "Thats Anty

---

* A Jarrow Colliery valuation of 1809 mentioned "84 Yards of Penning in Sledge Way", and this is thought to refer to a form of running-way or rail (Lewis 1970, pp.309,363).

** A blower is an outburst of gas or water, or both, through a crack in an underground wall or coal face, from a natural reservoir behind. In this case the blower began by blowing out mainly water, which was inconvenient but not dangerous. But when it stopped making a noise, gas predominated.

*** Nicholas Urwin was perhaps the man of that name, aged 58 and with eleven children, who was killed in the 1812 explosion when said to be working the inclined plane (Hodgson 1813, pp.36-7). Robert Stove was one of the survivors of the 1812 disaster but was killed in another explosion at Felling in 1813 (Sykes 1833, 2:76-7).

**** The gas had "backed the air", that is, it had advanced this distance and although lighter than air was beginning to fill the whole tunnel from roof to floor.

Errington!" He came running to the pit. I told him to call of [= for] John Straker, Viewer [= manager], Ralph Brown wasteman [= maintenance worker] and all hands. Emedtly the furnes was to [be] put out, and we were in great Danger.* Straker Came down and the westman went Emetly and put the furnice out, which was at the Roddney pit. Brown and Straker, Stoves and Urwin and I went in to see how it was going on. We herd it roring as the presser of Steem from a Large boiler. By this time the gas had backed the Air 6 yards out past where I was working. "We have seen plenty" - and we made al possable spead Out, to go to bank [= the surface]. The waste men had got the furnace safe out. We got Safe to bank and Mr Straker said I was worth my weight of gold. I had saved my own Life and Others and the Whole Colliery. N.B. It was 3 days before the Blower spent off, after which the work got on again.**

---

* The pit had a furnace in one shaft, apparently underground, in order to heat the out-going air and so increase the circulation. But the gas was also being drawn out of the mine and an explosion would occur if it reached the furnace-fire and was ignited.

** A.E. inserts at the beginning of this episode, and in the middle of it, a series of notes on the development of the Hollihill pit, one of the pits of Felling Colliery. The quoting of exact figures and dates suggests that he was drawing on a work diary for 1800-1801. The notes refer to a number of underground tunnels ('drifts') which had been excavated ('driven') in different directions from a small underground shaft ('staple') in which a winch ('gin') was employed. One of the tunnels ran to a district of entrance roads and coal-working galleries ('headways' and 'bords'). In the main entrance roads, along which the coal was to be brought out to the shaft in waggons, A.E. was busy laying waggonway. One tunnel was approaching the boundary of the coal worked by this particular pit, and was therefore near the area worked, or formerly worked, by a neighbouring pit (which took its name from Admiral Keppel, first lord of the Admiralty 1782, died 1786, a popular hero at Newcastle). Also this tunnel (or perhaps another - it is not quite clear which) was thought to be approaching some abandoned workings which might have water in them: hence borings through the coal at the head of the tunnel were being made to test for water.

"South drift from the Hollihill pit 14 fathum below the seam of high main coal. 374 yards to a Staple sunk 21 feet and a gin to draw the Coal. Aprile the 10, 1800. 140 yards of way from the

 <continued on the next page>

One month after this, on the Monday morning, Joseph Hunter, Overman [= shift foreman], and [a] Horskeeper and a fue boys went Down first. Nickel Urwin and Marton Greenar got in the Loop and I got on the Rope above.* When we had Decended 30 fathom, the Rope serged on the rowl [? roll] and we would drop [= could drop] [a further] 2 fathum.** I Called, "Hold! Stand fast to Look to the rowl!" We had to be slung to the pullie fraim. 80 men [were] at bank [= on the surface] [waiting to go down] and they pased us a Small Cord to Lash our selfs to [the rope with]. The Depeties [= district foremen] did so but I Could not Efect to do so. The overmen, feared of me falling, got 2 bottles of hay Lightened up [at the foot of the shaft] and their Jackets on top. We hung One hour and 20 minits before the rope was put wright on the rowl. William Anderson said, "Keep up thy hart, Anty!" In about 10 minutes they Lowsed [= loosed] the seesing [= seizing] and Lowered [us] to the Botum. I Could not stand [up, my] Leggs and thies was num. I striped off and Joseph Hunter Rubed my thies and Leggs with both

---

<continued from previous page>
botum of the Staple to the west, the Coal rising 2 inches in a yard. A Stone drift from the top of the Staple to Cut the Coal 118 yards to South West. Aprile the 17, 1800. 167 yards of way laid South and East down to winning headways from the Staple botum South."

"From the top of this bord it was droven down 627 yds to boundery, to South from East End of Huerth Church. West way at this time, West from 118, West do [?] 111. West and North East side of Admirel Keppell pit, Boarhole 60 yds of Soled Coal left. 27 Janury 1801. 133 yds West and boaring feared of Old pits above the High Felling, 2 of such pits was not down to the high main."

* Martin Greener, a deputy, was killed in the 1813 explosion at Felling (Sykes 1833, 2:76-7). A man of the same name, who worked at Felling as a youth c.1805, and who by 1841 had become resident viewer of Tyne Main Colliery, was probably his son (RCCM 1842, p.629).

** The winding rope, as it was being lowered down the pit, somehow jerked and slipped off the "rowl", and jammed.

hands until the Blood got Circalation. They [i.e., the others] was done the same with. Once more I returned Almity God sincear thanks for his Merciful providence over mee.

Hillpit fire and loss of life <34>

At this time the East board was driving East down, and 3 days before the Drumer was got, the bord fired [= exploded] on the Friday and one man [was] burnt. The bargom men [= bargain men, men working on specific tasks by contract] gave up on account of [the] fire and none [was] at work Setarday and Sund[ay] night.* On the Sunday night I was down and [also] John Clark, the Only two that was down that night. I assisted him, an Old man, to load the timber waggon. He got the horse and went Inn Alone. I had some repairs to make in the South drift and it was 2 hours when I went to see after him.** I met him near the Staple [= underground shaft]. He said, "Anthony, you must not Leave me this night for I have seean a Spiret". "Dont be feared, John", I said, "Let me see where you saw hit". We had to go 50 yards near the pit [and he said], "This is the plase. It was a man with [a] fearce Countinence runing at full Speed". I said, "Neal down, John !", and we Boath prayed to God to

---

* The episode described in this anecdote occurred some days before the episode recounted earlier involving 'The Drummer' <32>, but it apparently occurred in the same locality, even though A.E. now seems to refer to the "East bord" rather than the "South East bord". Gas which had accumulated in the gallery, or 'bord' exploded; and the specialist workers, who were working on a contract as 'bargain-men' and should have worked over the weekend preparing the district for further coal-getting, refused to work there. A.E. was braver - or more rash.

** At the weekend, maintenance work was being done: timber was being brought in to the district to provide the roof-supports required as the work advanced, and A.E. was repairing the way. The tunnel from the shaft was high enough to allow the timber-waggon to be pulled by a horse - A.E. implies that the colliery horse-keeper was later burned in the same locality, so perhaps he had gone in-bye to lead the horse back to the underground stable.

Preserve us from the Danger of the pit. At the same plase, as I was going in [again], I saw the Spiret with fearce Countnence [which] pasd me on the Left Side runing at full speed. I said, "O God, be with me and this man" - which [= who] was then [with]in hearing. We kept together [and at] ½ past One James Bell, horskeeper, and Joseph Hunter, Overman [= shift foreman], Came down. John Clark told him what he had seen. He would belive nothing what he said. We Asended the pit [and went home, but] at 8 oclock I went [back] over to the pit. Men was coming up, and they got out of the Corf [= coal basket] and Jumped Over 2 [other] Corvs and roon home. This was James Bell and Thos Craggs. As soon as I saw James Bell, [I recognised] the Spiret was the same. He [had been burned and] Died after 20 hours.* Craggs recovered. Thus the Providence of God made me meet at the top of the pit the same [man whose spirit I saw below]. He Left a wife and 7 Children, al girls. On the Tuesday 2 depeties [= foremen] was to go to this same board, but William Hunter went alone. The whole Air of the pit [was] forced by a brick wall to 4 feet off the Coal in frunt. He was on the Left hand side [where] the Air turned to the righthand. [Suddenly a blower blew out.] His candle was 3 yards back [from the blower], he Clicked his Candle and there was no fire. And he run [out] and told all to run for their Life. Near 9 oclock in the morning I [had] just got on the pit top when he asended the pit, and his fase was al

---

* Miners involved in underground explosions often received extensive superficial burns which, while not immediately disabling, made the sufferers frantic with pain. Horrifying descriptions of the injuries thus inflicted were given in a medical treatise on burns written by a Newcastle doctor (Kentish 1797,1817). Miners were generally given medical attention in their own homes since the few hospitals on Tyneside were located in the large towns, some distance away from the mining villages. In the 1790s and 1800s probably as many as several hundred North East miners died from burns (or from the septicaemia which often set in as a result of uncouth medical treatment). In this instance, the burned men were sent up the shaft in a corf because they were too badly injured and shocked to ride the normal way. James Bell, aged 40, died on 29.4.1803 (Heworth burial register).

Cut by the blower. This [blower] was what turned out to be the Drumer.*

Grand state queston <35>

Advertised in all the Newspapers of England, Ireland and Scotland - the queston, what was the real valliue of the 3 Eunited Countries ? This was 3 weeks advertsed. I was working alone and had my thought on that. That Evning I made a Statment and Directed it to the Printer, Solomon Hotchen, and he might give it a triell. I Sined the Adress Anthony Errington, Mackenick, Felling. After 10 days [the] Lunden papers [said] the grand queston was Solved by a Meckenick of Newcastle upon Tyne.** This sooen spred and I was Called the greatest Scholar of England and Could not walk 5 yards on the Streat wouthout being requested to [join] ther Company. Questons [were] proposed to me, was it done by Algebra or Problems or by Disimals? I declined answering theas questons. For 2 or 3 weeks I Could not keep Clear of drink. I had to Leave off going to Newcastle for a month. [Then] I had to go to High Frier Streat, to [my] Brother in Law, and there was 6 Shop keepers Come to the house Inquiring for Me.

---

* Apparently this blower started with gas, then turned to largely water, before resuming a major outburst of gas.

** The reference to "3 Eunited Countries" and the later reference to "the Population ... taken that Spring" seem to date this episode to 1801, the date of the Union of Great Britain and Ireland and of the first census. But if the reference to the printer "Solomon Hotchon" indicates Solomon Hodgson, printer of the 'Newcastle Chronicle' (and no doubt agent for the national newspapers), it is awkward that he died in April 1800. However, the paper continued with the name of the printer given as "S.Hodgson" - actually Solomon's widow, Sarah - and A.E. may not have known this, or remembered it when writing twenty years later. Unfortunately I cannot trace the notices A.E. refers to, either in the 'Newcastle Chronicle' for 1800-1801, or in various 1801 London papers (I have searched the columns of minor English news in the 'Morning Chronicle' for all 1801 and in the 'Morning Post' and 'Lloyds Evening Post' for the first half of that year, also the index to 'The Times'). I may have overlooked the item, or it may be elsewhere in the London papers - or in a paper I have not seen.

Brother and I went with them to the bay and barrell. They wished me to tell [them] how I solved the queston. [Seeing] that I was the greatest Scholer in England, they would propose a queston. There was 4 questons probosed, [but] I declined answering, [as] I could not put off my time there. One [question] was, had God ann equall? The second, what weight of Preser of Air on men´s body? The others [being] in Algebra and Dismells, I would not start to please them. [But] there was a Stout women there called Nanny Hare who Carried flesh [= butcher meat]. She Called for a gill of Ale and hearing their questons, [said,] "I will propose a queston that none of all ye Seven Can Answer, that is, What Never was, nor never will bee?" I went to the backyard to Make warter. There was a Small room [ad]joining, and she went into this and gave me a Hare [= hair], you ges from where, and Said, that never was Streght nor Never would bee.* The glases [were] bet that I could not solve that queston. [But] I Produced the hare and they Owned [up] and paid the glases. I stepped [aside] and gave the women a glass unpercieved.

[The] real queston was that the Population had been taken that Spring, of men, women and Children, so many millions, [and what was their value?]. [My answer was that] the goverment of Great Briton vallieued Ever[y] one at £50 0s 0d a piece. And that if we the Peepell did not defend it [= Britain], it would belong to Some Others. That was the real valliew of the 3 Countries.**

---

* The "hare" was presumably a pubic hair, and the name of the stout and waggish female may have been spelled either way, both names being known on Tyneside.

** It is a pity that A.E.'s account of his successful answer is so condensed as to be obscure. However, it appears that the answer took the form of a hearty patriotic slogan reflecting the popular and official viewpoint during the current long war with France. A.E.'s source for the average value of each individual has not been traced, but a local newspaper had referred to earlier estimates, by Petty and Davenant, of the average "expense" of each individual (Newcastle Chronicle, 25.10.1800).

Felling colliery 1803 - the bone setter <36>

Anthony Errington was going to diner along the Summerhouse row [when] Jane Robson Came out and Exclaimed, "O Less [= ? alas], My bairns [= child´s] neck is out!" I jumped into the house and the Child [being] on the floor, I took hold of [its neck] and plased the underparts between my nees and got the Edge of my hand under the Jaw and puling gentely, the bone went in with "knack". The Child Cried. The Mother [was] siting, I gave her the Child and shee fainted and I had to hold her and the Child and Call for help, and thus 3 women Came. I got my hands weshed [and the mother then explained]. The Cause [of the injury] was [that] the Busem [= broom] had fallen betwen the dor and [the] hing[e], and She Coming down Stairs and [carrying] the Child, she [tripped over the broom and] fell and the Child also. Thus providence saved her [? the child´s] life. I then got the name of Cleaver [= bone-setter] Anty.

Bad money <37>

At Felling I was paid Evry 14 days. I had to go to the Ofice for the pay note, then I went to the Sighn of the Shakespear to get Change. I had 3 mens Money and my own. I Called for £2 of Change [for] pound notes. Robert Miller, Landlord, brought the 2 papers [= packets]. I opened one and hee went away. On opening the Other, they was soverings [inside] instead of Shillings. I instently rung the bell [and when] he Came I told him he had given bad money. He flew into a rage of pashon. I said. "Look hear!", and on seeing the Soverings he said, "I beg pardon, it is a grand Mistake". He brought on to the table One quort of hot Ale and Brandy and the Landlady said, "Well done, honest Anty!".

Newcastle date <38>

My Niboor and I, when at Felling, went to Newcastle and bought Beef 2d per lb below the morning price at 9 Oclock. We Left it for the Cart on Monday. On going on the End of Newcastle bridge I had to go to [the] Necesary [= public latrine]. The man got up at my right hand and there was One of the Dealers, a Woman, Came in and said, "With Your Leave, Jentlemen, them that is Neadfull Canot be mindfull". She sat down near me and droped, [then] took my Shart [= shirt] Lap and wipered, and Said, "Your foaks will wesh before ours. Goodnight!" I was greaved at her treating me soo. The trick was in the Newspapers, I was Called a farmer, the Name not nowen [= known]. I Called this a Newcastle date.*

Proverbe <39>

"He that digeth a pit for his Neighbour shall fall unto it himself."**

Before the Felling [pit] holed the waste [= abandoned underground area], the One pumping ingine was double Shift. George Bailey [was] Plugman in One [shift], Mathew Wilson [in] the other, and they Changed shift fortnitly. Mark Stevenson was fire man to George

---

* A later anecdote <72> refers to an episode at the town of Alnwick (locally pronounced Ann-ick) which A.E. calls an 'Anick date' (i.e. anecdote). So A.E. calls this a 'Newcastle date'. The newspaper reference has not been traced but the Newcastle papers did contain anecdotes about tricks played on simple rustics (e.g. Newcastle Chronicle, 28.3.1801). However, A.E.'s anecdote parallels (if the episode did not actually engender) a local legend. A recent popular booklet asserts that in a public convenience in the Old Cattle Market men and women used to sit on opposite sides of a partition, back to back; and that "one farmer suffered the indignity of having his shirt tail pulled under the partition by a 'lady' and used as a toilet roll" (Graham 1986, p.30). As late as 1845, Gateshead had "no public necessaries" (Health of Towns Commission, 1848, report of D.B.Reid, p.100), hence A.E.'s use of the one on the bridge.

** A variant on Ecclesiastes 10.8, Psalms 7.16, or Proverbs 28.10.

Bailey. Mr Straker came down in the Thursday morning and, near me, said, "Mark, what was the Mater Last week ? The Steem was bloing off [from the engine] at 2 Clock". "I cannot tell, sir", [Mark replied], "Wee stoped 1½ hours, and sometimes 2 hours, to get a bit rest, then we start again".* [After that] the Orders was that the Ingine was to work Single Shift. George Douglas was working wright [in charge of the engines] and he and George Bailey was provocked at Mark teling Mr Straker [what he did]. [So] the two formed a plot for Mark. A fue days after, the Plug had to be put into the Londery box [landry-box = device on pump to distribute water] and all Locks stoped in the pumps. I was working near [where] Mark was whealing his Coals in, when Bailey Called him. They Stoped the Ingine and I followed Mark to se what was wrong. Bailey was up Stairs and George Douglas was at the plug fraim. Douglas said [to Mark], "Thou has to put thy thumb on heare", [and he put his own thumb] upon the Scogen [scoggan = Newcomen engine valve-opener] of the Injection. Baily [from upstairs] seeing the thumb on [the scoggan] through the Crift [= ? cleft] of the floore, nocked the prop out and the Scoggon took Duglases thumb off by [the] second joint. He Exclaimed - it was the rong man! I withdrow without being seen by him and on that I kept it to myself for fear of their Envie. This I saw and the Inicient Man Escaped the plot.

Pill docter <40>

I was returning home of the pay Friday Evning, very Stormey from the North East [with] drifting snow. On going [along] alone, I herd a moan and Stood and Lisent. I found a Man in the railway guter, Overhead Snow Lying [across him] and water under the snow. I got him Out and took him to Padick Hall. He dweled in

---

* The simple-minded fireman told the manager that the enginemen were spinning out the work in order to keep the pump going double shift.

Gateshead and had got two much. I weshed him and got super and then to bed. [Next day] he had to look for his box at a publick [house] where he had bean in the Morning. The frost was very Strong in the Morning. Thus I was [an] instrument in the hand of God to Save [the] Life [of] Dr Anderson, [the pill doctor].*

Asalt at Mr Barkers closet <41>

I dwelled at Felling and on the Setorday was at Newcastle and in the Flesh Market bought 3 Stone of Beef and put in to Mr Barkers Closet at 11 Clock AM. [But when] I Called for the bag at 3 Oclock he had sent it away in a Mistake with the Jasment [= Jesmond, a Newcastle suburb] Market Cart. He sent 2 girls for the bag and took me into a Small room and [gave me] a pint of Ale and [the] London pabers. Theire was 4 young men Come into the room. One Said, "Your the man that killed my Brother, you Bu....., I will kill you". The kee was in the door inside, I Locked the door [with the] 4 inside. He took the Pore [= poker] and the Other the fire Shovell. I kept the pore [away] with my Left hand and wrenched it from him. [As for] the Other, I Struck at the Shovel neor his hand. And I gave the first a Tempeler [= blow to the temples], and the Other, [by] whealing round, gave him a Back hander over the Jaw and Cheak. He droped. The Other [= another] was Striking at My Side. I tosed him throw the window and the Other jumped Out after him. 2 Sergents [were] Called and the 4 was taken into Custody, [the] Jergents [sergeants] Frank Jackson and Miller. The window was valued 16s 0d damage. I said they were under a Mistake, they took me for Bilton. Jackson New me, My Name was Errington, and it was Critical to now [= know] One from the Other. I have seen him restore pease when a froction [= friction]

---

* For a later rescue by A.E. of the same Dr Anderson, see anecdote <73>.

took plase more than Once.* They Begged parden. The girls Came with my beef from Jesment, and the Sargents, Landlord and I had a glas Each. They would not request me to atend the Court on the Monday.** They would State the case and Said Like was a bad Work. I parted friends and returned home. Bilton had fought with a Mason and the Mason died after the fight.*** Bilton Left the Miller the west Contrey and went South and was no more herd of. This was the Shaddow taken for [the] real man.

One Month after I was at Barkers and he again gave me the papers and took me to the same room again and I had a Look at the New window. Shortly after he brought [in] a Old man, a Collier belonging to Willington.**** He Brought him a pint of bear and Said, "Now you will sit Comfortable heare". I red the pabers to myself, [but] Shortly after there Came in 4 young men [who] brought their pint and glas with them. They [laid] there pint on the same table with the old mans, and they Soon insulted the Old Man, Called him Bad names and Drunk of his pint. I had my Eye on them. They puled off his hat and wigg and One of them Struck him twise in the briest [= breast]. He burst into tears. I Laid the paper down and Asked whie they cused the Old man so. The answer was, "Had [= hold]

---

* Presumably this refers to Sergeant Jackson.

** This may mean that there was no court case: perhaps the men bribed the sergeants to be released. But this seemingly did not happen the next time.

*** Bilton, the Mason, and the Miller were apparently prize-fighters. There were certainly prize fights in Newcastle, but I have not adequately identified these names in 'Boxiana', perhaps because this contemporary work seldom touches on pugilism outside the Home Counties. A prize fighter named Bitton was superannuated by 1820, one named Mason was a minor fighter still alive in 1821, a notable pugilist was West Country Dick, and 'milling' - hence the Miller - was the sport's in-word for 'boxing' (Egan 1818-24, 1:470,2:502,3:440,554,556). A.E. seems acquainted with newspaper reports of the sport.

**** Willington was then a mining village near Wallsend.

thy toung or we will give thou the same". I jumped up and gave him a tempeler, I took the other heals out, the 3rd [I gave] a blow in the Stomick, the Other [went] out the doore and [was] Stoped by the Brower and brought back. The Old man [being] Still in tears, the Landlord Called and the same Sergents Came. The Old man Said, had I not bean there, "they would [have] killed me". They were handed off to the Ketty [kitty = gaol] for the asolt. The Old Man said I was to Come that day fortnet and I would be paid for my kindness. I did so, and there was 32 men and women Came which Ofered [me] Money, but [this] was not taken. [Instead] I would take a glass or two with the Menn. The Sergents Came to see me and there was 5s for me for the asalt, and the Magistrates Said it was a Manly action to Save the Old man. They was bound Over to [keep the] peace for 12 month with the old man and had all Expence to pay. Thus I was rewarded 10 fold by the friends and Many Others for Standing up in defence of the Aged.

The peneypie jack in Newcastle <42>

In the Three Bulls Heads I [was] in Company with a taylor and there was 6 men Come in from work. They would buy 2 pies a peace if hee [i.e. the pie jack] would Change a pound note. Change was bad to get and he had [only] 6 shillings. I Lent him 14 shillings which much Oblidged the Men. He Said, "O Lovly, I will now [= know] thee again". 14 days after, I met him. [He said] I was to have a pie and he gave me the Money Borrowed. 6 months after, [when] I had Bought 3 Stone [of] Beef Cheap and my pocket [was] nearly ashore, we met on the Streat.* [He said] I was to have a pie. "Go in", he said. I went with him and I told him I was nearly ashore. He said, "Lovly, there is ½ Crown for thee, more if thee says what". I saw him no more for 8 years [because] he was transported

---

* In money terms, 'ashore' is the opposite of 'afloat'.

to Vandemons Land for 7 years. [The one day] I was in the Old custom house, Sandhill, when in Comes Jack with his basket. He Looked [at me] and said, "Lovly!". I took out the hauf Crown and One peney for a pye, and laid [them] down. He sat down and shed tears and said, "Lovely, one hundred times and more I have thought, when far away, should Lovely be living, I will [be] hauf Crown to the good". The Brewer hearing what pased, brought a full quort on the table to ingoy our Selves, which we did.

Holing the waste [= abandoned section of seam] with loss of life <43>

On the 27 March, Good Friday, 1807, the Discovery pit was working to the west and the Crain was 13 piller[s] up.* George Hunter and George Riddly was in this board [= coal-getting gallery], 5 boards north of the Crain, Hunter on the right and Ridley on the Left. Betwean 10 and 11 Oclock Riddly holed with the pick [into the waste and water rushed into the bord].** He fled before the warter. Hunter stoped in the Cornr of the board. Hee was drowned.*** John Car was a puter [= putter, youth pushing trams or small waggons from the face] and Andrew Anderson was at [the] bailing water plase, [working] to keep the Air Cours Clear, and they was Drowned [too]. [Other men having been cut off by the water were unable to reach the shaft,

---

* Pillars of coal, dividing working place from working place (that is, bord from bord), were left to support the roof.

** The waste, a section of the seam abandoned because the coal had been worked out (sometimes by a neighbouring or earlier colliery), provided a space, despite roof falls, which often filled up with water. To hole into the waste could therefore be one of the most disastrous occurrences in coalmining, for if the waste contained water, the new workings might be flooded out within a few hours, or even minutes, trapping numbers of miners.

*** It later transpires that Ridley was drowned, not Hunter. In whatever way punctuation is supplied, the text is in error - it was presumably Hunter who fled and Ridley who stayed.

so] Robert Brown, overman [= shift foreman], William Hunter and I decended the pit. When we got near the dyke [= break in the seam], the warter was 6 ft breth, 1 ft 6 in depth at 4 inch fall in a yard. The way was all weshed away, and we Could not hold a foot. Near One a clock the feeder [= underground stream] abated and we got in, to se for men and boys surviving. Al got out safe, this 3 wanting [= excepted], and a 4th was at the top of a pump, [still cut off and with] Little Hope for his Life. On the Setorday the warter abated [further]. On the Setorday Evening all hands [were] on search for the [missing men], Ridley and Carr. [I thought that] one of thos men at the pump top might be alive [although] the One at the bailing [point] was gone. Edward Rodgers and I went Emedtly to Mr Straker who saw the hope and Ordered 18 men to go with him directly. They had 18 yards of Standing [= airway partition] to take out and to ther Joy they found the One [man], Ralph Hall, alive but very week, and took him home. At 2 Oclock Sunday morning he Came round again, and was ordered by Mr Brandling to go no more down the pit. At 4 Oclock Easter Sunday, we, Robert Brown, William Hunter and Anthony Errington, decended to search for the two bodies wanting. There was a board full of water to the roof. I striped off and swim in with a peace [of] Chauk in my mouth and made a Chauk [mark] 25 yards in [where] I could go no further. On the Sunday Night we pumped this warter out and found Riddly 4 yds bye [= past] the Chaulk. Car was found under a fall of Stone on the Tuesday. The 2 was buoried at Heworth. I Lowered Riddly into the grave.*

On the Thursday following I had to have 2 depeties [= shift foremen] down the pit with me, [so] I went near 3 Oclock PM to James Trumble who dwelled next door [to Riddly]. I went up to the house. When I Entered the

---

* In fact, all three victims of the indundation were buried at Heworth, George Ridley, aged 35, John Carr, aged 18, and Andrew Henderson (sic) (Heworth burial register).

house - Ridly dwelled up Stairs - I cast my Eyes up on the right hand doore. And I saw Riddlies Spirit turn into his Own door. Trumble was [standing] on the Left hand down stars when he said, "Anty, I have seen Riddly!" I told him to say nothing and I told him to go for Edwd Rodgers, Overman [= shift foreman]. We counseled [together] and I told them the sole Could not rest [because] there was money hidden. I had to get Other men that night [to do my work]. I went with Rodgers to Mr Robson, the Agent, and told him [what had happened]. Who held up his hand and said, good God, was it so? We then had to Send for the widdow and Douters. He kept them on tauk [for] 1¼ hours [while] Rodgers and Trumble surched and found Coin in 3 diferint plases, [worth] near £9 in the whole. Which was given to Mr Robson to give her when she might want. His Sole was seen no more. May he rest in peace.

Bad company <44>

I was at the Freemason Lodge, Gateshead, [one night], and after leaveing the Lodge, William Ray and I [were] in Company.* Ray dwelled at Felling Shore and I dwelled in a Cotage where Lees factory is at preasent, and we came together to the Friers Goose Steath [staithe = coal-loading quay]. On pasing that, Ray said, "Now Devell, if thou bee a devell, Come onto the road and let me see thee!" On this, I Left the foot path and went on to the head rig [= field furrow], and said, "O God, deliver me from the Enemy!" Wee walked the Length of the field. Ray had to go threw a Stile and I Left him and made my way Alone throw the wood [where] their was a guter [= ditch] 3 feet brith to Step over. I was feeling with my foot for the Edge when a ball of fire came down the wood. I jumped the

---

* In 1794 the lodge was held at the Blue Bell Inn, Bridge Street, and in 1811 at the Goat Inn, Bottle Bank (information from Mr M.Henry).

guter by the Light given and I got Safe home at 20 Minets past 12 Oclock. At 6 Oclock in the Morning I was going to work when I was Called on to Stop by William Ray, who asked what time I got home. I told him and he said, "Dam the Devell to Hell, it was 4 Oclock when I got home and I hardly have a whole bone in my body. I have bean over tree tops and thorn hedges, and all the Close that I had on is torn to raggs". His wife told me a Short time after that she burnt them in the Oven. He was an unbelever of Athiest Cast of mind. I [was] never more in his Company. I returned God thanks for my preservation from the poor [= power] of the Enemy.

Singlor battle in Newcastle <45>

On the Setarday morning the Wife and I went to Newcastle Market and in the Bucher Market Bought Beef and Muton. The Bucher gave us a Stake and went with us to the [eating] house in Dowrie Lane. He Left us, and we got what Satified [us] and 2 pieces [of meat were] Left and Bread. The house was Boxed, One table and Bench on Each Side. Each seat would take fowre [= four] and wee took the right hand Seat. I set the Basket on the window, [then] 2 of Jqure [? Squire] Hawks Blacksmiths Came in. I ofered them the plate and warm meet, which they acepted, then got Change and parted their money. They was siting near the Wife. On this a Farmer and a Miller with his dusty Close Came to sit in the Other Side [of] the table. The Farmer sat down [but] the Miller said, "I will not sit a Side such durtie things", [meaning the smiths]. The Smith said, "The Clean Man and [his] wife had no folt with us siting near them", and [that] he was a dustie Miler that would daub any One. The Miler said, "Thow little dirty thing, I will kick thie arse". The Smith said, "I dar [= dare] fite thee". They went into the Lane, the Brower on One Side and I on the Other to kep the Craud back. [They were] a very unequall pair, the Smith 7 stone, the Miller 18. They had 3 meetings.

The Smith soon had him blind and he had to be Led into the house. The Landlord and the Company [said] the Miler was the first igresor, where on he was to pay One gallon of Ale to the Company. He paid the gallon, and hauf an Our after he would fite him again. The Brower and I went out [again] and kept the Craud back. They had Other 3 meetings. The Miler never hit the Smith. The Last meeting the Miller wis strideling [= straddling] and Lowing [= allowing] the Smith in behind him, throw betwean his Legs. And [the Smith] made the Blood flie out of his Nose and One Eye [was] shut. The Miler gave in and was Led into the house and pa'd the Second gallon of Ale. Before the Seccond fite, I said to the Smith, "Thee is Mr Hawkses man, hear, there is 5s 0d for thee, keep up thy hart". 4 Munths after I had to Order some Waggon wheals at Mr Hawkses. Near 12 Clock PM the Clark and I was going to have a Litle Chease and Bread. On Entering [the public house], the Smith was [there] with his Fauther and [the father's other] 6 Soons, and [said], "Heere is the Man that lent me 5s when I fit [= fought] the great Miller". The Elder Brother Jumped up and Called for a glass of brandy. The Clark said, no, not yet, the smith was to go with us to Mr Hawks. The Clark had got the whole detail and told Mr Hawks. He [said] "Good man. For Backing My Servint, I will give thee 5s 0d. Go and get your Lowens [= allowances]." I got no ferther that day. The afternoon pased away with [the] farther and 7 sons. [Afterwards] they Left Hawkses and went to Hull river.*

---

* This episode is of uncertain date and "the Wife" might be either A.E.'s first wife, in which case it occurred between 1798 and 1809, or his second wife, in which case it occurred c.1820. However, it is more likely that a wife went shopping in Newcastle from the first wife's home in Felling than from the second wife's home at Backworth. The iron foundry of the Hawks was at Gateshead, but "Hawks and Co., anchorsmiths, Sandgate" (Directory 1801, p.21) presumably indicates a Newcastle branch of the firm. Wherever they worked, the smiths whom A.E. encountered apparently lived in Newcastle, and on a Saturday midday they were returning home with their pay (they "parted their money"). The "Mr Hawks" was presumably one of the three brothers who were partners, George (1766-1820), William (1772-1807), or John (1774-1830).

Death of wife <46>

During the time I dwelt in the Cotage my wife took very ill. I at that time was Making a railway at Sudick for Mr Wake and I was sent for to go home in all haste.* She Could not speek when I got home and she Deprted Life on the 9 March, 1809.** Thus I was left [with] 2 Suns and 2 daughters to Lament the Loss of a Mother and honest partner in Life. I shortly after Left the Felling work and went to Benwell to work. I dwelled at Felling but was at Lodgines at Benwell.***

A mason saved <47>

On the 21 of October, 1809, returning home on the Setorday, I met with Padison Solsby, Brother trade man.**** We had 2 quarts of Ale and at 2 Oclock parted. I made my Marketing [= market purchases] on the road to the Bridge [and put them] in a Small bag. When on Newcastle Bridge I saw 2 Masons repairing three Chemelys of a house in Gateshead, the house was Opisite the Blue Bell. I thought they was in great danger. I [left the bridge and] walked in the midle of the Streat. [As] I Looked up, the rope broke. One

---

* 'Sudick' is Southwick, a village five miles east of Felling and three miles west of Sunderland. The railway built there was not a colliery railway, and why A.E. was so employed is not explained.

** The burial of Ann Errington, aged 40, "of Kirkwhelpington, daughter of Fenwick and Ann Hymers" (i.e. Hindmarsh), is recorded in the Heworth Chapel register.

*** A.E. had lived at Felling since birth, but he now begins to roam. His first move was to Benwell, a village on the western outskirts of Newcastle, where there was another colliery owned by the Brandlings, the owners of Felling. However, since he returns to Felling after a few months, this seems to have been only a temporary transfer arranged by his employer, and the final, permanent break with Felling Colliery had yet to come.

**** The 'brother tradesman' was presumably a waggonway wright. There may have been in existence a trade association, or 'brothering', as it was often called in northern Britain.

Clung by the Slate, the other Came head first down. I droped my bag and plased my self [in position], One fot in the Chanel, the Other on the flag [= flag-stones]. [He dropped on to me, and] My back to his back, he Carried me down. Mr Liddle of Ravensworth, [a] young gentleman, [was going by] and his Servent Riding behind him.* He jumped off the horse and Lifted me up and Called for a Cergen [= surgeon]. And said, "The Nobelest action that Ever was done by Men!" We was taken to the Hauf Moon and he demanded the frunt room.** Wee boath Striped and the Cergen Exemined us and [found] no Bones Broke. [But] on the Sunday I found there was a Leader of my Shoulder was put [out]. On the Monday, Joseph Wood, Edge tool maker for the Hawks, and Charles How of Newcastle, Made a Collection of 13s·4d on the Streat for the Man that had Saved the Mans life. Which was laid upon the table for Me, [but] I Could not take money for Devine Providence, I was not my Own Master. I presented it to the Poor of the Parish of Gatshead, and I was to be a freeman of Gateshead. Squire Liddle paid 5s to the Cergen and treat[ed] Each of us with One glas of brandy, and he said the Man [saved], Joseph Dalton, Mason, was to treat me One glas more, which was done. One of Hawkes Smiths had my bag and all I had bought was safe in hit. I made home haveing Sore Shoulders from the fall which wore off in a fue days.

Divine Providence <48>

When at Benwell I lodged at Fenum Lodge [with] Mrs

---

* Ravensworth Castle, not far from Gateshead, was the family seat of the Liddells, a wealthy and influential family of landowners and coalowners. If the "young gentleman" was actually of the Ravensworth line and not one of the other Liddells living around Gateshead, he can be identified. At this date, Henry Thomas, the oldest of seven sons of the eighth baronet, was aged thirteen and a pupil at Eton.

** Although some distance from the action, the Half Moon was a superior establishment, an old coaching inn, and no doubt used to serving gentlemen.

Bowman, a widdow. Their was a man, a Baptist preacher, Stoped [= ? lodged] there and I saw him [once] give her 18s 0d which he said he had got that day for preachen at Newcastle. On the Sunday their was a girl that had Lived Servint with me Called to see the femely.* [She said] her Sister was Married to a Yorkshire weever - the Father and 2 Sons Masters and Imployed 2 Jurnimen. [But] the Fauther took off to [undertake] preaching, Leaving the wife and 2 Children. From the Driscription given [by the girl], he was the Same [man who had stayed at Benwell]. Near 12 months after, [when I had left Benwell], I met the preacher and the Landlady [together], and we went to the inn and had One pint of Ale. In his absence, She Let me see the Widing ring they had bought that day. I tould her she was not to do soo, he had a wife and 2 Children. [But she would not listen.] Her husband had bean Stuart for the Earl of Carlile and was Left £400 cash, the Intrest of 4000 in the funds. She had One Sun, [a] Farmer near Morpeth, his name Anthony Bowman, and I sent word to him to Look Sharp after the Mother.** [About this time] I was accosted by a Man Selling Blankets [who asked me] to by a pair, which I did. [Since he came from Yorkshire] I asked him if he new Such a man [as the preacher]. The Answar was, quite well and whe[re] he dwelled. I told him of the deacait he was doing, who said, "I will wright to his wife to Come down". She Came the week following. And the Sun Came the Same day the wife and Children Came. I was driven in Speret to go to the house also. Mrs Bowmen Called him [i.e. the preacher] "Satton begon!" [= "Satan be gone!"], and the Sun turned him Out never to come there no more. I told him [i.e. the son] how I discovered the Imposter, who Said the Providence of God had gided mee. He took her home to be more Safe and I saw thease [people] no more.

---

* The reference to Sunday implies that the call was made on A.E.'s family at Felling.

** Morpeth is a market town 15 miles north of Newcastle.

Recalled from Benwell <49>

The Hollihill pit had worked sometime as Land Sale for John Grase.* John Straker and John Brandling had 500 Chalder [= Newcastle coal-measure of 53 cwt.] of Coals [stored] at the pit.** The railway was taken up but theas Coals was Sold to go to Scotland. [So] I had to begin on the Monday morning with 12 hands and Lay down 345 yards of railway, which was done on the Wedensday at 4 Oclock and 3 Chaldern of Coals [were] Shiped that Evening. The Wager of 20 guineas that the Ship Should have the Coals on the Wedensday was bet [and won] by Straker. We was rewarded with One hauf Barrel of Ale and Cheas and Bread that Evening.

John Pit railway - April the 18 1811 <50>

951 yds of Cast Iron railway [ran from John Pit] to Padockhall turn. [It was a] great day at John Pit [when] the first Coals [were] taken to the River on the 8 of October, 1810, with 21 guns firing on the Balest Hill.*** My farthr was playing on the fiddle Down the waggonway on the top of the first waggon.**** When 50 yard from the Steath [= staithe, coal-loading

---

* The paragraph heading is in fact the only reference in the text to A.E.'s return from Benwell to Felling. John Grace was a partner with the Brandlings in Felling Colliery (Hodgson 1813, p.6). Probably Hollihill pit was now landsale because it was nearly worked out.

** In the manuscript there is no punctuation between the names of the individuals and possibly all three owned both the colliery and the stored coals.

*** The date given in the title is not that given in the text below, and the significance of the former is not explained. It was perhaps the date A.E. left Felling <54> and if so may be attached to the wrong title. The Ballast Hills, lying along the river bank, were formed from ballast dumped by incoming coal ships - a Ballast Hill near Felling Shore was shown on the 1856 OS map.

**** Robert Errington, now aged 79, had been retired for nearly ten years, but having worked in so many of the Felling pits in his day he was invited to attend the official opening of this new one.

quay], One of the guns burst in 3 peaces, and One peace Cleared My Farthers head [by] 12 inches. But on this I was Charged not to tell him by Mr Brandling, it would Damp his spirits. He had to Dine with the Company at Felling Eunicorn Inn.* One John Hunter had My Fauthers fiddle in Charge while the Company was at Dinner and he was playing to please the Company down stairs, [when] Mr Straker Came down, seased the fiddle and Broke it over a Chair top. After he was informed what he had done, he Sent to Felling Hall for his [own] fiddle. The fiddle [belonging to my father] Lay broke 12 years and then Straker got it repaired, the cos[t] £1 4s [for] repairing.** Thus I got My Fauther Safe home and so Ended the great day at the first Coals from the John Pit.

Hardware shop in The Side, Newcastle <51>

I, Anthony, entered the Shop [one day] to buie a gimlet and handsaw file. 2 Brothers, Turnbull ther name, [owned the shop].*** One of the brothers was up stairs poorly and at near 3 Oclock PM the Shop filled of people in a fue minets. I was requested to Look to a Little, and for 1½ hours had to wait. [Then] I was going to Leave, having got the gimlet and file, [when] a Cart Stoped at the door, and the Owner Came in. He said [to Mr Turnbull], "Will you take [your] Money?" And Mr Turnbull said to me, "Stop!". He paid [Mr Turnbull] £15 for goods he had [had] Cridet for and I had to witness the payment. He Ordered £15 more goods to be redey on the Tuesday following. 2 months after I Called at the Shop to buie a pair of Pliers. They

---

* The Unicorn Inn was on Felling Shore (1856 OS map).

** If it is correct that the fiddle was mended in 1822, this must have been to oblige A.E., since his father was by then dead.

*** The Side is a street in Newcastle leading down to the Quay. Newcastle directories of 1787 and 1811 do not list an ironmonger named Turnbull, but the 1833 directory lists Mary Turnbull, saddler and ironmonger, in Union Street (Directory 1833).

was Laid on the Counter [for me] when a young Jentleman Came in and said these was Just what hee wanted. Heare the Second time the Shop filled and [the] two [? of us were left] waiting and I keeping an eye on the whole with [? him] for 1½ hours. The Brother that was poorly said, "Good God, what a man this is!" [At last] the People got Served and I was going away when the [same] Cart Stoped [again] and the Owner Stoped mee. The Marchant paid £15 0s 0d [again, which] I had to witness. [Then] Mr Turnbul and hee and I went to the Pack Horse and had a glass a peace. He travelled Northumberland and the borders of Scotland. Mr Turnbull told him of the Shop filling twice, who said, "God! Providence was with mee!"

Temtation to bee reserecton man <52>

One Setorday I was in Newcastle and with [my] Brother in Law at the Cock, head of the Side. Their was 2 well dressed Men of the Scotch Dilect [there], who would treat us with Sperits. They said they had Come from the South and they wanted a Strong man to assist One of them to take there trunk down to the queyside. [They said my] Brother might go home and I would do very well. At ½ past 9 I went with the 2 men. I asked [them] where the trunk was, who said, "We will soon be their". On the North side of St Nicklos Church yard was [a] wood Railling, and on going down the railing 3 of the rails was out, and [this way] the 2 men got into the Church yard.* I made Warter and Considered that they might Murder mee and send me off to Edinbrough.** Where on I made the best of my way

---

* "In the year 1761, St Nicholas' churchyard was inclosed with a brick wall, with rails upon it" (Brand 1789, 1:271. The rails are clearly shown in the engraving in Mackenzie 1825, opposite p.235.

** The Edinburgh medical school was notorious at this period for its alleged use of stolen corpses. The term 'resurrection man' first appeared in the 1780s (OED); in 1792 there was a riot in Berwick-on-Tweed directed against 'anatomists' (Sykes 1833, 1:361); and on 28.11.1801 the 'Morning Chronicle' had an item headed 'Resurrection Men'.

to Newcastle bridge and thence got home. I had been in a grave before.

A man prevented from hanging <53>

On the 21 Feburary, 1811, I had bean at South Shealds on buisness for the Colliery. Returning home, on Hebron Fell there was a man came riding hard and Said, "My Lad, Can you run? Their is a man going to hang himself. Follow mee!" He turned off the road and I followed him, having run 3/4 of a mile, and at the uper End of the field, the Servint man of George Worker was up the tree. He had the rope fast to the tree when the horsman got hold of him and Said, "Come down, thou must not do that this time!" I Led the horse and he Led the man to the Farmhouse and delivired him to his Master with Orders to watch him. His name was Lightfoot. The horse man and I at Low Huerth had One pint of Ale. He told me he was Ellington the Miller. He was Called a wise man [in the district]. We parted at Huerth Bridge End, and he said, "You will not be [easily] fritned?" and I said, "No". I had not gone 100 yds when the Lightnings flew in all Sides of me, yet I took no hurt. I saw him Once in Gateshead after, who said, "How did you get home?", and he Lafed and said he was jesting. 4 years after the [servant] man did hang himself.

George Hawkes drunk <54>

On Setorday night I was returning from Gateshead. It had Snowed 2 hours, [but] the moon got up and 50 yds from Gateshead Streat, [on the] Sunderland road, I saw a man Lying Covered up by the Snow. I hollowed him and he answered, and then I got him up and Led him to Heworth. The Bad Language he had would have made any man Leave him on the road. We got to Hueth and had One pint of Ale [at an inn], and I left him. 14 days After, he and the Wife was in the Bucher Market. He said he was own me a pint and I went with them, and She, the Wife, did not now [= know] what to say bad on me, I should [have] fetched him home, [not left him at the inn]. That Evening, as I was returning home, he had got two mutch and Cowped [= tipped] the Cart at the Robers Corner, the wife in the Cart, near dark.* I had to set to and get them gathered Out of the gutter. She had then Nothing to say. Shortly after, I met them [again] in Gateshead. He had [just] denied My Fauther of Carrieing his Market Poak [= sack] in his Cart to Felling. I told him One good turn deserved another. But [they said] that was not their wish, to do any good [turn] to a papist. Where on I said, "I will see thee Lying in the guter before thou dies." 3 weeks after, I found them in the same ditch [at] Robus Corner. I Stood and Looked. They was no worse but boath drunk and the Limer [of the cart] Broke. I gethered them up the Second time. She said, "I thought a papist would not have done so to a protestint!" I Left them to go home.

Shifting to Percy Main <55>

Mr Strakers fine promises to Anthony proved false. He gave George Douglas 24s per week and would give Anthony no more than 18s 0d, [although] he was no beter workman. On this I Left [Felling] and went and Ingaged [myself] at Percy Main on April 24, 1811, and Started [there].** Against my mind I Left. I was Leaveing my Aged fauther, and none to help him but 2 Sisters. Strakers Coldness to Me was bace. Mr Robson the Stuert wished me to stop in Mr Brandlings Imployment and Said Straker was Out of his head.***

---

* 'Robbers Corner' was the curve on the Sunderland road near where the entrance lodge of Park House then stood (Manders 1975, p.130).

** Percy Main, a colliery established only in 1799, gave its name (which referred to the Percys, Dukes of Northumberland) to a new colliery village, located about six miles east of Newcastle.

*** A.E. had worked for the Brandlings for nearly 20 years, and his father had worked for them for 30 years before that.

But I had recieved Arles [= engagement money] there [i.e. at Percy Main], for I was [committed] to go. My Enemies Reported that I had gone to a place where I Could not make a Livelyhood for My Famely. [But] the first pay for 14 days [was] £3 1s 0d. And on the 17 May My Son Robert Started [at Percy Main], and Reiseved wages [for] 9 days at 1s 0d per day from John Taylor - he was waiting on [= assisting] the Banksman [= workman at the pit top]. When done with that, he was Imployed with his Fauther.

My son very ill <56>

The Howdon pit Holed into the Percy pit and the Water [which came through] had bean Standing 8 years, which was Like as much Vitrol.* He and I on the 24 of July, 1812, had to work on the way half Leg depth of Warter for 12 hours. He got weshed and Brickfast and went by Water to Felling. When he got there he took Very ill and got the Doctor, who Said he was in a Fever. I went to bed and Could get no warmness inn My feet and Leggs. [So] I got One gill of Brandy and Weshed them and wraped them Close up in Flannen [= flannel].** I went to bed again and I got 3 hours Sleep. And [being] all over sweting, I was quite well on Sunday Morning. On the Sunday night I went to work at 6 Clock and got home at 10 on the Monday and [found] a Letor saying that Robert was Ill. I made no delay but got the Steamboat up and was at my Fauthers at 12 Clock.*** The Doctor came at the time and I told him

---

* Howdon, a village near Percy Main, gave its name to one of the two Percy Main Colliery pits.

** From what follows it appears that the brandy was applied externally - although A.E.'s brother George had survived fevers by internal application <23>.

*** A.E.'s recollection must be wrong. The first steamboat only came into use on the Tyne in 1814, two years after this episode (GM 1814, p.344). The stated date was a Friday, therefore A.E. slept not only for three daylight hours during Saturday but also through Saturday night, before returning to work on Sunday evening.

what was the Cause [of the fever] and what I had done to Myself. Who Called for a bottle [of brandy] deirict and weshed him with it. This being done I had to Leave him to go to work that night at 6 Clock. My Son was 14 weeks down of the Fever. The Soles of his Feet nearly came off. [Afterwards] I bound him to Mr Hueison of Newcastle to be a Painter, plumer, and guilder for 7 years, which he Served Faithfull to his Master and the Master was faithful to him and made him Compliat [his] yeares.* God be with him wheare Ever he goes.

1813 Howden pit blasted <57>

On the Whitsunday Night 44 men decended the pit. The Overman [= shift foreman] George Cooper and 2 depeties [= district foremen] went to the East Croscent [? crossing = air crossing], [where] their was a blower [of gas] in One of the places 18 yd up [and a] bretich [= brattice or partition] [? to conduct the] air. The Stone had fallen and put the door down and they put the door up [again]. [But meanwhile] the fool [= foul] air Came a Long the back headwas [= headways = passages] and Came onto the main way to the Shaft.** We had got on to work when I saw the Candle just at the Firing point.*** There was 6 Wastemen [= maintenance workers] with me and I shouted, "Put out the Lows [= candles]!" Andrew Bell said, "Lord have Mercy upon us, we are all dead men!" I asumed the Comand and Said to Ritcherd Martan, "Run and put out

---

* Probably "John Hewitson, glazier and painter, head of the Side" (Directory 1811, p.48).

** The collapse of an air door had disrupted the ventilation of the pit. As a result, the gas from the blower had accumulated and was moving the wrong way, towards the place where A.E. and his gang were working.

*** The presence and the proportion of gas in the air are shown by changes in the shape and colour of a naked flame. A.E. saw from his candle flame that the quantity of gas around him was almost sufficient for an explosion to occur.

all Lights and Stir up the furnice and Make the flame go up the Shaft." He had to Tell the Furnasman and then got the Pore [= poker], broke [up] the fire and the gass fired [in the shaft] and went up into the high Regens.* It blasted 6 times and the whole Contery [around] was at the pit [soon]. A man and horse [went] off to Walls End for Mr Buddle.** None would Come down unto he came. ½ past 9 he Came and Samiell Cooper [with him] and the Swet was Droping off them. Ther was 2 Doors to Percy pit and the Scale [= leaking air] Came [though] to Howden. One man was going throw the door [but] I told him he was to Stop and to keep his Light in, which he did. One man went in in the Dark to the Overman who would not Come Out. After 2 hours waiting in Silince the Man Came [back] from the Furnace and told us that all was well out by [= out-bye = at the shaft]. We got Lights and begun work. When Mr Buddle Came, he was astonished at

---

* The gas was deliberately exploded as it passed over the ventilation furnace at the foot of the shaft. Because it was night time and there were few men in the mine, the furnace fire was low, and hence the ventilation was poor and could not be trusted by itself to carry the gas away to the surface. Therefore the furnace fire had to be stirred up (most furnaces were so arranged that the foul air passed up a bye-shaft and avoided direct contact with the fire). This explanation attempts to make sense of a very curious episode. It may be that an explosion in the shaft was less dangerous than one in the workings, but it was still a very risky solution - and not perhaps an altogether necessary one.

** John Buddle, whose home was at Wallsend, a village near Howdon, was the most eminent colliery viewer of this period. The overall manager and supervising engineer of many mining concerns in the North East, he was also consulted professionally by coalowners in all the other British coalfields. He acted for many years as the Secretary of the 'Vend' - the famous cartel of North East mines selling coal on the London market - and he was one of the founder members of the Society for Preventing Accidents in Coal Mines, the body whose activities led to Davy's invention of the safety-lamp. Buddle gave evidence on the subject of the coal trade and on the subject of accidents in mines before several parliamentary committees. A man of some education and culture (unlike most of the earlier colliery engineers), he was a friend, or at least a respected acquaintance, of many of the great coalowners of the North East. He was generally respected by the miners because of his willingness to go underground and attend personally to problems there.

seeing the Candles burning and [this after] the Pit had Blasted, and said, "Men, how is this?" Andrew Bell [said], "Mr Buddle, had it not bean for Anty Errington, wee would have bean all dead men. He scried [= observed] it Just at the Firing point, and you know whates is heare, 3 yds off - 18 boards and 16 piller of Creeper Waste bared off by a bord end Stoping!"* Mr Buddle said, "Good God, what Providence! Such a thing for the pit to blast 6 times and no man hurt! The Like has not bean in the Anells of history". Mr Buddle proceded further Inn and on returning brought the Overman out with him - he would not believe it was so. And Every man that was in the Pit was Ordered to be at the Ofice at 12 Oclock on the Monday. Mr Buddle had all the Owners of the Colliery preasant. Andrew Bell was Call[ed] in first and Said to the Owners that Anthony Errington was the first that Scried it Just when it was going to Fire. "He gave the Comand, put out the Lows, and Ordered Ritcherd Martan to the Furnice. He saved all Our lives." Anthony was [then] Called on. Mr. Buddle said, "Anthony, before the Owners of the Colliery, thee hath saved thy Own Life and all the Men and Horses and the Colliery. I will Confer the Overmans place on thee." I thanked him [but said] there was young men in the Colliery Coming up in Exptacion of Such places, I had served 7 years a trade, all I wanted was to Live by my trade, and [he was] to give his favor to such [as I had mentioned]. He Answered and said, "Owners of this Colliery, did ever man Make a more Pathitick speach toward his fellow Man?" He, Mr Buddle, shook hand with me and said, "Anthony, thee

---

* A creeper waste was part of a seam that had had to be abandoned, because an advancing pressure wave of the strata above, operating on the pillars of coal supporting the roof, had caused the floor to rise or 'creep' and the pillars to crumble. Collieries at this period were often troubled by 'creeps' during the process of 'robbing' the pillars (that is, reducing their size as the working retreated) and Buddle was much concerned with the problem. A creeper waste, being unventilated, was often full of gas; and in this instance Andrew Bell argues that if the gas from the blower had ignited, combustion would have spread to the creeper waste and the pit would have blown up.

shalt live by thie trade all the days of thie Life." Which promise he fulfiled to his death.*

Society esteblished at Percy Main for the betar ventalation of the mine <58>

[The society was] to consist of Overmen [= shift foremen], depeties [= district foremen], wastemen [= maintenance workers], hewers [= coalface workers] and Others, Boath pits being in a dangrus State. I itinded one Meeting and I saw Clearly that the Overmen downed Each one that Spoke.** [But] I wished to be herd [and I said this]. Suposing a fire to take plase [underground] and [the explosion to] take away the [air] doors betwen the pits, [then] the whole of the Air [is] taken off the Intearier of the mine. The After damp [= asphyxiating gas produced by an explosion], Left to stand [without ventilation], produses death to thease [= those men] [caught] in it. [So] I proposed then to Make the Improvement which I saw they was short of. I was Shouted at by the Overmen in great derison, what did I now [= know] ?. [But] Silence being insisted on by the hewers [who said] I was to speek and not to be interupted, I Stated that there should be a Valve door [fitted underground], whose Frame was to be cut inn the soled Coal or Stone on Each side and with Strong hinges at the top. This door [was] not to be in use [but] kept open by a prop at One side, and thus to be redey to

---

* Since Buddle did not die till 1843, and since all the other evidence suggests that A.E.'s account was written before this date, the final statement may be only a rhetorical flourish. As for the offer of the post of overman, a very responsible post, it must be doubted whether Buddle would ever have considered that A.E. had had sufficient experience of mining technology outside his own trade to have made such an offer. Therefore this statement is most probably rhetorical embroidery.

** A.E. means that the overmen, who should have been the technical brains of the mine, treated the suggestions put forward by the inferior workmen with great contempt.

keep the Air in its Course after a fire.* The folowing morning Mr Buddle Came down. One of the hueers told him how the Society was Conducted and that Anthony Errington [had] made a Preposition which was nocked down by the Overmen. He, Mr Buddle, Came [to me] and Inquired at my Self what it was corning [= ? concerning]. I told him the aforesaid, who Said, "Go with me to the shaft [of] Howden pit". He marked the plase and the Door was Made and a fraim Cut in [the] top and side so that a blast would keep it up Each [= either] way it Came. The Same like door and fraim being done at Percy pit.

Saved by a dog

<59>

[This] being done, we was all that atended the Society treat [= treated] with a super at North Shealds. To frame Love and Friendship toward Each Others welfare - this was Mr Buddles intent in doing so. We parted all friendly and returned home. On the road, there was a Strange Dog came to me and walked Close to mee. I said, "Poor fellow, has thee lost thie Master ?" I gave him a peace [of] Spice and he kept Close to me. I dwelled at Howden, and [on my way there] had no company but the Dog which I called Tom. On going along the foot [path, with a] faint Moon in the Last quarter, near 1 Clock, 2 men got up from some bushes

---

* A.E.'s argument here is sound, and the suggestion which he claims to have made was a sensible one, although perhaps not an original one. The high mortality associated with gas explosions in mines was normally due, not to the direct effects of blast, but to the generation in the explosion of an asphyxiating gas ('after-damp', a carbonic acid compound gas), which, because of the blowing down of the air doors and the consequent disruption of the ventilation, drifted through the workings killing all in its path. A.E. therefore suggested the installation of additional air doors, hinged at the top, which, being normally propped up, would not resist and be torn down by a blast, but would automatically fall shut thereafter. In this way normal ventilation would be resumed immediately after the explosion and the gas would be rendered harmless by mixing with the incoming air. Doors of this type were in fact adopted in several collieries at about this date.

and said, was I going to Howden? I answered, yes. [They said], "Have you any mony?". I Answered, I had 1s 6d. [One said], "B..... your Eyes, Lets have your Money!". [I said], "Sees him!", and Tom the dog downed him. And I minted [= ?] to Strike at the Other Man, who, seeing the Dog upon his Marrow [= mate], Evaded me to rescue the Other. I then run [away] One hundred yds and I hard a foot Coming after me. I swated [= squatted] down thinking it was the men, [but] it was the Dog which Licked my Left hand and Left Blood on my hand. It Set off from me and Incercled me as I walked home Over ditch and hedge and it kept 20 yds distent. On reaching home, I gave him super and he Lay down.

The dog stolen <60>

[The dog stayed with me, but one day as] my second Son, [who was] at School, was tiring the Dog out in a string at dinertime, one Gorden, a Shipwright, seeing the Dog, took it from My boy. I got home from work at 2 Clock PM and [found] the boy in tears. [So I said] I would go very soon and get the Dog again. Gorden was a Stranger to Me and of a fiting Black Carrecter. I feared him, [so] that [when] I went to a publick house [to find him], I sent an Old Man for Gorden. He Came and I told him I had Sent for him to Drink first, which he did. [Then] I asked him by what Athorety he had to take the Dog from My Boy. He flew in a rage and was for Nocking my Eyes out. I Called on a Constable to take Gorden and we would go Direct to A Magistrate and the Dog with us. The Constable, a taylor by trade, took him in the Kings name as preasner. When he found this, he Carmed himself and Said, "B..... the Dog, I will go and fech hit". Which he did, and Delivered it in the String to [the] Constable and said he would have Nothing More to do with it. Where on [? we had] 2 quarts of Ale and [I had] 1s 6d to [the] Constable to pay - I paid the bear [beer] and he [had] lost $\frac{1}{4}$ of a day. The Constable spent the Shilling [? on more ale]. The boy Came and got the dog home. Thus wee parted friends. On the Setorday I took the Dog to My Fauthers and there [it] was a gardien to him and Sister. [Its] ferther history will be related here after.* Thus Once More Divine Providence Sent the gardien to rescue me from the Robers.

Fire in Howden pit
up the dike [= strata slip] to the west (61)

Some Men [were] burnt and One [had] his Leg broke [in this fire]. And One boy fell down the pit in Coming up. That Night Sevral Men went to Make all wright again. A Short time after, the pit fired again and the board [= coal-working gallery] took fire, the Coal and timber. No warter [being] near [at] hand, the Overman [= shift foreman], John Oliver, got a ships Gun and Balls, [in order] to fire up the board. [But] Mr Buddle Coming before the gun had gone down, Led warter and the fire ingine was used [instead] and put the fire Out. The gun was not used for fear of ingering the Stopings [= ventilation partitions]. [However] there was a Blower in the North headwas [= access passages] which burst down the Blue Stone [? blue-metal = form of shale]. This Blower Came out of the poast [stone] [= sandstone] and hised like a Serpent. Mr Budel [decided] to trie the utilety of the gun [with this blower]. Mr Buddle requested as many as had Any desire to See the Experiment tryed to go [to the place]. A Musket was brought and I, by [using] a Long Lat [= lath], Set the Bloer on Fire which [was] 8 feet above the Coalhead.** The fire was 2 yds Long and 2 feet thick, when the Musket Load[ed]

---

* Nothing further about the dog appears in the account as we have it. The "second son at School" was Anthony Fenwick, aged 10-12.

** If this really means that the blower was located eight feet above the top of the coal seam, then the height of the headway was at least 12-14 feet, presumably to allow for horse transport.

with Ball fired, and it put the Blower Out. [It was] tried a second time [and the result was] the Same, and a third time. And we was satisfied that the Cannon would have put the fire Out had it bean used at the board.*

Boddily fear of six men in the same way up the dike (62)

John Marchel and I was partners, and on the Setorday at 2 Clock PM we was Shifting the Crain 2 piller[s]. We had bean at work some time before they [i.e. the others] Came. We hard [a noise] as a tram [= coal rolley] runing, and Marshel went to See who was working. [But] he Could See no One. He Came and told me he Could not find them. I said, "Never Mind, Let us have Our Own work done". These 6 Men Came to Start [work] and before they Started they [too] herd the tram runing up One board and down the Other. [So] they Came to us to see if we had bean runing a tram. The answer was, "No, not us". They said the pit was ha[u]nted and they Durst not start. They went for Samiell Cooper, the Viewer [= manager], and he came with Bible and Prayer Book, and Acused us - what had we bean fritening the Men for? [We said] we had not done soo. We had to Leave off [work] and go with them 3 bords [= working places] to the South, and then there was Other 3 [bords] to the Barrier [of coal]. We was all to go to prayer when Marshel herd the tram again and Exclaimed, "D..., there its!". When the Sound Came there Longer and Stronger and Nothing to See they was sure there was going to be a Hevey Misfortune. All in Consternation, they would not stop [in the pit]. I said, "If you all go away, Marshel and I will stop. I have no feare on me, we will have owr work done first". I said, "Can none of you

---

* Accounts of Buddle's life refer to his experiments with percussion as a fire extinguisher at about this time. He was carrying out these experiments on behalf of the newly founded Society for Preventing Accidents in Coal Mines.

Percieve what it is?". No, they could se Nothing. I said, "Samie, they Are Loading the Rollies in Jarrow Corperation way and that [is] what we hear".* Sammie Cooper gave a great Shout - "Whe was aw telling that tee? [= who was I telling that to?]" I made Answer and Said he might go to Jarrow and Inquire and he would find it to be the Case. He said N[e]ither me nor no Man Could Make him believe that we Could hear throw One hundred yds of Solid Coal. [But] he went and found as I had Said. Wee got Our work done after they left us and got safe home. The way was to bee on [i.e. working] On the Monday, and [on] Sunday Night Overmen and Depeties [= shift and district foremen] [in] full muster Came [in] and Marshel and I was there. The way Started [up] and at 4 Oclock Mr Buddle came and [also] Cooper - he had told [Mr Buddle] what had transpired. Mr Buddle Came to Me and said, "Let us sit down". Being seated, he Said, "Anthony, thee has Made a grand Discovery. Thee hath A Penetrating Mind Inn a Coal Mine. Thee hath done a good dead [= deed]. The pit would have bean Laid in had thee not made the Discovry". And he said I was a perticular man. I said the Wise men of Percy Main Called me a fool - what did I know? [But] I judged them not. I was poss[ess]ed with a Lively Faith in the Providence of God Over Mee, and I Cared not what the wourld Said onn mee. [Mr Buddle said], "There is sumthing I want [to know] - what put that in your head, Anthony?". I ansered, "Experience. I was at Felling Colliery when waste men and viewrs 2 Came from St Antons Colliery. Their coale was worked in the high main before the Felling [coal].** And these waste men was sure that the Felling [pit] would hole in a fue days [because] they herd the Pick and Mell [= hammer] and wedge

---

* Jarrow faces Percy Main across the River Tyne, and Jarrow Colliery worked the coal under the river next to that worked by Percy Main Colliery. 'Corporation way' was a main underground waggonway in Jarrow Colliery (NCRO, Easton Papers, 3, Dunn's work diary).

** St Anthony's Colliery was located across the Tyne from Felling Colliery.

Clear, [although Felling had] Left a 50 yds Barrier [of coal between its workings and those of the neighbouring colliery]. They Lined the pit [= measured up the workings] and Stoped the work 5 hours without Equenting Mr T Barns, [the viewer]. The following day Mr Barns and 2 depeties Came [down] at Night and Lined the pit [to survey how far the workings had advanced]. I had to go with him to assist, when [it was found that] the Shortest Length of board to Drive [= excavate] [was] 18 yds, [while] some [were] 20 [and others] 25 or 30-36, [which] was the farthest - [yet] still Leaving 50 yd of barrier. That is the fact, Mr Buddle." [He replied], "I am quite satisfied. Mr Barns would be Correct. Your statement is good. You and [your] partner must get your Lowens [= allowances], and Samie will pay it". Thus the hearing [of a noise] throu the Soled Coal could be herd 118 yds. Thus I Proved the hearing and Not seeing at this time.

A Scotch piper <63>

I dweled at Howden Row and the tide was [? low], [so] the wherry went from Howden near Six a Clock AM. I got with it to Newcastle. It was Cold on the warter. On arriving at Newcastle quey I went to the Low Crain house which was Open. We got sumthing to warm the Inside and a good fire on, and we Soon Came round. There was none in when I entered, [then] there was such a rush into the house 2 rooms was ful in a fue minets. Their was a Lamed Man Atracted my notice warming at the fire. I gave him a drink and he thanked me in the Scotish dilect. He said he had Comed from Morpeth that morning on a Corn Cart and was a Stranger in Newcastle. He had not One peney in his pocket. [His] hunger and necesety touched my hart and I got him sumthing to Eat. He had a bundle in a handkercher and after he got refreshed he opened out his bundle. And he had the small pipes and [was] a great prefichint [proficient] on the pipes. The house was full of Men, [so] I got a plate and gathered on the whole Company. And I got and Laid down on the table for him £1 7s 0d, which he took, praising God for his Merciful Providence over him that morning. [He was] Called John Maccie.

A Scotch fiddler <64>

I dwelled at Howden. 16 Men had bean at work from 1 Clock PM on Sonday, and at 6 Oclock Monday Morning we went to Howden publick [house] to have Beef Stakes and Bread. And then near 7 Clock their was a Scotch Lad Lame in boath feet and his Mother [came in]. They had bean 13 days in Newcastle and had poor Luck. He had the fiddle. I asked the Mother, had they had any brickfast ? [She replied], "No, nor not One peney in the pocket". Tuched with Compasion for the poor and Nedey, I got a plate and gethered 14 Shilings for them. They boath nealed down and returned thanks to God for his mercies to them. I saw them Once after, and 7 years after Saw the Lad, his mother was dead. He had a Orfin girl 8 or 9 yeares of Age which he Suported. And I never Saw him more.

Out from the 'teapot <65>

[I was] Drinking Tea at the Hare and Hounds, Newcastle quay, [some time] after Mr Thomas Dods, Viewer at Hebron Colliery, Lost his Life.* Mr Buddle Esteblished the Widdow with what is Called a Monadge, that was, to pay 5s or 10s per fortnet for Linen and wearing aparall.** On the pay Setarday I Called to

---

* Thomas Dodds was still alive in mid 1813 (NCRO, Buddle Papers, 6/11), and was not named as one of those killed in the explosion of 12.5.1814 at Hebburn Colliery (Sykes 1833, 2:85).

** A 'monadge' or 'manadge' (from French 'ménage') was a box club instituted by a shopkeeper, often a linen draper. Members of the club agreed to pay in regular small sums, and goods could then be

<continued on the next page>

pay my Own money and [that of] Other **two** [men]. On going to Mrs Dodds room there was 12 women geting tea, and I was [invited] to sit down and take a **Cup** with them. [But] it was Brandy out from the tea pot. I was astonished. I had to take Mrs Dodds Book, and take the Money for Each ones Name [i.e. collect the money and make the entries]. The whole Company was in Capable of walking alone, [so at] 3 Clock PM I Led Mrs Dods to the Hebron boat. And the wheriman [and I] Each [took] two [of the women] and we got the whole [of the] women and their Marketing in to the boat. I went with them to Hebron. I took Care of Mrs Dodds. I set her home, and gave her the book and Money and [de]parted for Percy Main.

Intrige of a woman <66>

2 wimen was going along the footpath toward North Shealds near Chirton [when they met me].* One said, "Come along with us and I will put a wife in your hand". I walked along with them, [but] I met a man who said, "Good day, [beware!]".** I was aware of sumthing that they was going to, it was to drink tea

---

<continued from the previous page>
obtained from the shopkeeper, either up to the value of the total sum individually saved to date, or, more often, partly on credit, that is, in excess of the sum saved. In the latter case, the debt was repaid over time through the club, the person who organised it on behalf of the shopkeeper and collected the money, and who chased up defaulting members, benefiting from a percentage. (The system survived into the present century, and in the 1920s my grandmother collected door-to-door for a North Shields draper). In the case of Mrs Dodds, she was apparently presented with a stock of goods in order to start the monadge. The in-payments of 5s or 10s per fortnight were larger sums than one might have expected from working men at this period.

* Chirton was then a hamlet on the west side of North Shields, not far from Percy Main.

** This very significant story is unfortunately told confusingly and has required some editorial tidying up. A.E. was a widower in his mid-thirties, but when two women offered him a wife, a stranger - who presumably knew the two women - told the rather simple A.E. to beware.

at the sign of the Wormouth Bridge. This womens father and Mother and Samiell Cooper [the overman, were] there. I was to have gone to work at 6 Oclock [but Samuel Cooper] was so pleased that I had Comed with Jane [that he said that] he would Excuse my going that night. And [Jane and I], we would Ingag Our Selves. On the Sunday, a fortnet after, her Brother Came and I was to go down to his house near Six Oclock at night, [when it would be] very Dark. I told him I would Come Shortly. I praid to God to gard Me in the Jurnee and back, [but] as I was going down in a Narrow pas I went with my breast against a gate poast and [it] gave me a Stunner. [I thought], "Thou art as an Angel of God who met Balom in the way!".* [However], I considered to go [on] and make No Stop, and I went. [At the house], the botels [were] brought on to the table to help our Self. In Comes Samiell Cooper, a widdower as myself, and was so glad to see me their, and nothing would serve but to have the wedding day apointed. Because I was not well from the Stun I got, I begd to be Excused, I was quite unwell.

I returned home and saw Jane two times after that, and then She Left home. I Called to see her and the Mother said she had gone to where she was sheoring [? sharing] [because] the Housekeeper was ill. In due time there was a Lad from the North [who came to Percy Main] wanting to brake [= operate] an Engine. [But] the Ingine was not redey, [so] he was to Labour to Me. And the first night he told me what a Suprise his Landlady had put him under. He [had] menchened one Jane [whom he had met in the North and who was] Lying [in] at her Aunts at Spittle, and she [i.e. the Landlady] had wished [= hushed] him down never to mension that to No One. I was kind to him and gave him hauf My super and I got the whole Secret [from him]. A short time [after], the Ovrseer from Tweedmouth Came and the Child was a boy, father was

---

* Numbers 22.26.

Samie.* He Sold the Child to Mr and Mrs Hogg to bring up for £25 0s 0d and his friends Reported I had bribed Jane to father on him. Where On One woman who New my femely went to See this Child, and she Said it was none of Antys Get, his Children had Strong bone in them. It was not mine.** I had [had] Nothing to do with the women. The man that I met who said good day, [on the day of the 'engagement' party], told Me to be aware [= beware], and I was well aware that She would not do for Mee. She was a Solders wife, he a deserter and then in Irland. The Child Lived 6 weeks and died, and the Brothers Sun went to see Mr Hogg to have the money returned. No money [was] returned - there was 5 in kindred biter Enemies.***

No more peace at Percy Main <67>

After this I had no more peace at Percymain. The way was to be Let afresh and I was Cut out by One of my Men, by [his] Making Sam a preasent of a Pig 18s 0d valliue.**** Thus I was decieved in my Agreement. I

---

* Spittal, a fishing village 60 miles north of Percy Main, was in the parish of Tweedmouth. A duty of the parochial Overseer of the Poor was to establish the paternity of bastard children born in the parish. The Tweedmouth parish register, which records the 1810-1811 baptisms of boys born to two unmarried women each called Jane, has no record of a similar occurrence in 1813-1814.

** A.E. denies again that he was the father of Jane's child. But he did have at least one illegitimate child. According to the register of St. Andrew's Roman Catholic chapel at Newcastle, "Die 12 Aug 1810 nata est ... Sarah, filia Ant. Errington et Joannae Richardson". Sarah appears later in the autobiography. A.E. had been a widower for nine months when this child was conceived. It seems unlikely that the "Joanna" of 1810 was the Jane of 1813-1814.

*** The point of the last remark, a late insertion, is obscure.

**** Waggonway wrights worked on contract, wholly or partly, and the overman saw to it that A.E.'s contract was not renewed. Waggonway wrights paid part of the expense of laying way out of their own pockets, but then had only to maintain the way to earn a bonus for every score of waggons using the way. A.E. lost his contract before the maintenance-bonus recouped his capital outlay.

had Laid 3 Inclines and found Nails and Candles for Score price which at the begining did not pay wages and I had to Sufer Loss. When I Left I was £15 0s in det for nails and Candles and the totell Loss was £30 0s. I went to Fawden [Colliery] and in One year Redeamed £15 0s.* The Next year I got all put Stright at Fawdon, [working] as Waggonwright. Thus I many a time Said the gate poast was an Angel in the way to deter me from the women. May God Judge betwean us, he is dead and I am Living. His Son George Cooper was Called the wisest man in Percy Main. It was he that shouted me down at the Society, and what did I Now [= know] ?

Starting at Fawdon Colliery <68>

On the 12 August, 1814, I was Ingaged to Fawdon [Colliery] as Waggon and Waggonway wright at 19s per week. I was at Lodgings 2 months, and on the 22 November [when] Begining to make 12 Chaldren Waggons at £1 4s per piece, I got a house and coals at Hensharber.** [My] Doughter An, 14 years of Age, [was my] houskeeper. [But] there was at that time a Fever raging and She took the Fever and died in 2 days. Which was a Loss greatly felt. She was Buried at Gosforth.*** I got a houskeeper that I had before at Felling. I was Car[e]ful and saved all I Could to pay dets and I got all paid.

Hungry milkmaid <69>

Before my daughter died we got One pint New milk in

---

* Fawdon was a colliery and a pit village about three miles NW of Newcastle.

** Hens Harbour was a row of houses at the east end of the village of Kenton (1863 OS map), the village before Fawdon, coming from Newcastle.

*** Her burial is not in the Gosforth parish register.

the Evening. [One day] I had got my Tea and was done when She Came with the Milk. I asked her to have a Cup of tea. She [said] she would, she had not got One bite that day. Wheare on doughter put more tea in, I Cut some bread and doughter put buter on. [I said] she was to make a good tea and Welcome. She said their was not a bite of Bread in the house and when she got home with the Milk they got Each a peney Roll and Milk. The femely Shortly after Left Kenton and I saw her no more for 13 years. [Then one day] I was going to South Shields and in the Sleck Row their was a Women Called after me, was my name Errington, and Anty? I said, yes. "Come heare, I want you!" I returned back and was [asked] to go into the [house]. What did you want with me, [I said]. Her husband was siting and Said, Come in. She said to her husband, "This is the man that Saved my Life when I was dieing of hunger". [She said] I was to have somthing to Eat and drink. I took [a] slice [of] Beef and Musterd and [a] glass [of] rum. Her husband Said she had told [him] 100 times of that, and I was quite welcome. Her name was Hope and she said she had not Changed her name for his name was Hope [too]. Thus I was rewarded after 13 years.

A vission <70>

[One day] I went to Newcastle and bought the Market[ing] [= supply of food] and Sent it home. There was 2 Brothers called Howeys and John Stewert, a Catholic, and we had a drink of Ale and pased away the Evening.* It was 11 Clock and wee 4 set off home to Kenton [by] the foot path from the Kowgate along the west end of the Moor.** On this path the Brothers was

---

* John Stewart was probably the man of that name who in 1821 was a proxy godfather for the Thelwalls of Heworth, a couple for whom in 1807 A.E. was also a godfather (CRS 1936, pp.262,297).

** The 'Moor' is the Town Moor, an open tract to the NW of Newcastle and along A.E.'s route to Kenton. 'Kowgate', on the west side of the Moor, now a built-up suburb, was then a few houses beside the Ponteland road.

ir deapth [of] Ergiment Concornin the Isralites at the pasage of Jurden. One said they were Murderers, the Other the Contrary.* John Stuert and I Left them and Walked onn. We was within One hundred yds of the North Hedge [when] we Met 2 women [with] ho[o]ds on, One Man before, 4 bearers and pall and Corps, one Man behind. Clos to this was following the same on the foot path. We stood off on One Side and Counted 43 Corps. We walked forward and the Last 2 or 3 came swiming over the hedge. We followed the Last and they all Esended Over the hedge to the West. We walked up to the Brothers who was Still in argiment. We Said to Each other that they had not seen the Vission and we kept it from them. We thought to ourselves that it was a forerunner of Some great Misfortune. 4 days after this, Raulph Jackson, Ingine Wright at Fawden, relates the Vision that Stewart had seen and Asked me what I thought on hit. I Studied and said I was Arm in Arm with Stuart at the time. He was astonished how I Could keep secret such a thing. His thought was the same as Ours, the fore Noledge of a great Misfortune.

Heaton pit holed the Jesment waste and [? this happened under] the toon moore [= Town Moor], [which] Debered [= debarred] 50 and upwards of geting out of the pit, and they had Life for some months. [This was] on the 23 June and they was there 6 month.** One

---

* Joshua, chapters 3 and 4. But A.E. seems to have confused the story of the crossing of the Red Sea with that of the crossing of the Jordan.

** The disaster at Heaton Colliery (on the NE outskirts of Newcastle and some of whose workings may have been almost under the Town Moor) occurred on the 3 May, 1815 (not the date given by A.E., 23 June, a Friday, which is therefore also not the date of his Sunday visit). Some 75 workmen were trapped by an inrush of water from the waste of a nearby abandoned mine at Jesmond. The bodies were not recovered until nine months after the disaster. It was popularly supposed that the victims had remained alive for part of this time. In fact, all were dead within a few hours of the first inrush of water, being either drowned, or asphyxiated by the gases which were driven into the workings where the miners had taken refuge.

Sunday Morning I got up at 4 Clock. I could not rest and went Out to walk. I was Led by the Speret to Heaton pit. I was Standing Looking at the pumping Engine when the Engine Man said, had I any kindred in the pit? I answered, no. He said Mr Budle and the wastemen was down, he was Expecting them [up] Every minet. He sined me to Stop. They shortly after Came up very miry and exasted. They Requested some Spiret and revived and walked home. Mr Budle said, had I any friends below? I Answered, no, and he asked the reason of me [being] their. I told him I was not my own Master. I was Led by the Devine Speret to that Interview. [Then] I returned home to the Femely.*

The second vission on Newcastle moore <71>

I had bean at Felling and Left after diner and going throw Newcastle Called at the Fawdon house.** The Same John Stuart was their and was going away. I desired him to stop and take [a] share of My pint, [then] I would go with him. We set off and the sun set when [we were] at the Entering on to the Moor. And Crosing to the grandstand we Saw a great Company of Men Coming up the North side and Said to Each Other that they had bean at a buriel at Gosforth.*** We walked hard to fall in with them. [But] soon they was out of speeking with, and we Stood and waited to See what road they took. They Esended over a hill and went out of Sight. Wee waited to See them rise the

---

* If A.E. walked to Heaton from Kenton, he walked some three miles. But he says that it was a Sunday and that he "returned home to the Femely". This must mean either that he walked to Heaton from Felling, and back again, about four miles each way, or else that his household at Fawdon/Kenton included not only a housekeeper but also some of his children.

** Why A.E. calls this public house in Newcastle the 'Fawdon house' is not explained.

*** The Grandstand on the Town Moor, erected in 1800, stood where Kenton Road now turns off Grandstand Road (Middlebrook 1950, p.151).

Other hill. We thought they was long and we went forward to See. Nothing their. We went up to the Kowgate and Inquired after the men, their was none Come there. This was a Short time after the Heaton men was got [out of the pit] and burri[d]. It pleased God in his Devine Provedence to reward us with the figgure of Soles as Men in the Dawn of the Day. We said, God give rest to these Sowls.

The Alenwick date <72>

Ralph Lawson, Banksmen [= pit-top chargeman] at Fawdon, had a Sumens for the Court of Alnwick. He Came to me [for] their was no time to be Lost, and I went with him to the Master. And there was upwards of £3 0s 0d [to be paid] and they advanced [him] £4 10s and [told him] he had to go [immediately] and he would have time for the 5 Clock Coach to Morpeth.* He set off and got all settled and returned. At the Kowhill fair [that year] their was 16 of us Fawden men in Company, and the Master and a Schoolmaster of Dinington came in and was herty [with us].** He [said] he was glad to see the best of the Colliery there. [It was agreed] we was to sing a Song Each or pay one Glass. 2 Songs was sung and [it was] the Schoolmaster next. He Said he Could not Sing but would relate an Anick date [= anecdote]. On which Lawson, heiring, Exclaimed, "Away with your Anick dates, Let us have a Newcastle One!". On which the whole [company] burst into Laftor. The Schoolmaster up and Left the Company much ofended. The [school] Master could not refrain from the Anick date, [but] it was Newcastle date that evening.

---

* Alnwick, pronounced locally 'Ann-ick', is a market town 30 miles north of Newcastle; Morpeth lies between. Apparently Lawson had been sued for a debt.

** Annual fairs were held at Cow Hill on the Town Moor on 1 August and 18 October (Brand 1789, p.437). Dinnington, north of Fawdon, was then a village with a small landsale mine.

Saving a mans life <73>

At Fawdon I was returning home on a verry Stormy night and [with] Drifting Snow on the road side. [There] was a foot path 4 feet breth [but] this was Covered up. I was on the Middle of the road [when] I hard a Moaning and Sought to find [the source of] that. [I found it and] I got the man out of the Snow and [he was] benumbed of Cold. I had 100 yards to get him to My home and I warmed [him] by the fire. He was the same Docter Anderson that I had Once before [rescued]. When he Came round and Looked at mee, he Exclaimed, "Good God, this is Anty! Thee has saved me the 2 time. I did not now [= know] who it was till now. God be praised for his Providence over Me or [by] the morning I would [have] bean dead." I got super [but] he could Eat Little and got to bed. [In the morning] he got brickfast and the road being blown up [with snow], he had to make for the turnpike road to Newcastle. This Doctor Anderson [was] alwas travelling with pills and [was] subgect to get two much. He was an honest man and well respected in [the] Counties of Durham and Northumberland. He Dwelled in Gatshead to his death.

Leaving Fawdon <74>

I had got the New waggons finished and was repairing and on the pay night I was Informed that my wages would be the nex pay 3s 0d per day, that was 1s per week off. On the Monday I got a Masage [= message] to go down to Gosforth, there was One [person there] wanted me. When I got there it was Mr Thos King, the Viewer for Mr Brandling. He had bought 10 waggons at Murton and he wanted a waggon wright. On naming that to Mr Wm Brandling, he said, where was the Erringtons? "I demand One of the Erringtons, my old servents." Thus I went down and was ingaged, giving 11 days Notice to quit. [It was] 3 weeks before I got a house. I got 20s per week and was working at Gosforth when Bell and Brandling bought Cocklodge Colliery and On the first of August, 1817, Every man was tret with Strong bear, with Colors flying and Cannon firing.* I made the New frames and was fuley imployed for the first year. I then had to go to Cocks Lodge to work.

Married again <75>

My Fauther Died when I was at Cocks Lodge, at the age of 86.** I then was intent on Marrieing and Our Nibour said, "Sur[e]ly you will give Mary the forst offer?" [so] I said, "Marrie, wilt thou have me?" She Answered, Shee would see what her Mother said. [It was] 3 Clock Setorday and [she would] go direct and se what answer [she got]. She returned on the Sunday night. They was agreable, whereon wee got Married.***

---

* When employed at Felling some fifteen or more years earlier, A.E. had been sent to work, apparently for a short period of time, at Gosforth, then a landsale colliery but owned by the Brandlings, his employers at Felling <**29**>. In 1808 a new waggonway from East Kenton Colliery to Wallsend gave the Gosforth coals access to the river. In 1809 the Brandlings opened a mine at Coxlodge, between Kenton and Gosforth, which was thereafter known as Kenton and Coxlodge Colliery, and which used the line to the river (NCRO, Buddle Papers, 3/19). In 1817 they joined in a partnership to buy up the older concern, East Kenton Colliery, which they then closed down (Lee 1951, pp.169-70). A.E. is slightly mistaken.

** Robert Errington died on 1.4.1818, a few days short of his 86th birthday (Heworth burial register).

*** A businesslike courtship on both sides. Why Mary had the right to the first offer is not revealed, but can perhaps be guessed. Mary's maiden name was Pearson: she may have been A.E.'s housekeeper at Gosforth/Coxlodge and she was probably a Felling woman aged 27. Although "our Nibour" probably indicates a Felling neighbour, the fact that Mary visited her mother overnight suggests that, if her mother still lived at Felling, then the offer was made elsewhere, presumably at Gosforth/Coxlodge. The episode occurred on a Saturday and Sunday, and at one time A.E. had regularly visited Felling on those days, but this episode may confirm that his family had joined him before his father died. No record of A.E.'s second marriage has been traced, but Mary had her first child baptised on 29.2.1819. If the marriage took place after July 1818, A.E. married her not only because she was (probably) his housekeeper and knew his children but also because she was pregnant.

There was One Richard Marshel had to Leave Little Gosforth and go to assist to put up High Heweth Ingine, on which I was Ordered down to Little Gosforth to the Charge of what was wanted. He had bean 3 weeks at Hueth and I met him at One a Clock at the Barras Bridge. He had his marketing and [was] going to Dean Houses.* I stoped and told him what had accured. [But] On the Sunday morning a Shave maker, John Brown, Called and [said] he was going to Longbenton with Shaves [= woodworker´s edge tools] [and I went with him]. When we got to the burn [= brook] the water had taken away the plank and we had to go down to the railway bridge. [There we saw] Marshell, and Brakemen, Banksmen, 2 Depeties and one blacksmith, all at work Cleaning out some pipes that [had] fured up. That morning [i.e. the morning of the day of the previous meeting with Marshall], before I went to Newcastle, [I had] asked the Brakmen concerning [these pipes and was told that] they would go to our pay Setorday. [But] it was Marshels pay Setorday. I told Marshel I would not be his ´beat need` and took my tools away on the Monday.** Mr King Coming on the Tuesday said I was to go to Gosforth. I said, no, I would have a hearing before Mr Brandling. [But Mr

---

* "Dean Houses" stood in Jesmond Dene, "lower down... above a stone quarry" and the houses were "inhabited by labourers" (Mackenzie 1825, 2:470). They were owned, at least at later date, by the Brandlings (Welford 1879, p.18).

** A.E.'s account of why he left Gosforth is far from clear. It would appear, however, that he caught the man Marshall participating in , perhaps in charge of, certain work which, in view of the time and place, he considered he should have been involved in himself - and for which he would of course have been paid. The expression 'beat need' is a dialect term meaning 'last resource, stand-in, stooge'. "Little Gosforth" is presumably modern South Gosforth, and the burn the Ouse Burn, lying between there and Longbenton. The final reference to Heaton, NE of Newcastle, seems to indicate that A.E. instantly headed from Gosforth, via Heaton, to Backworth, further east, his next place of employment. One thing seems clear from this episode - A.E. was becoming more crotchety as he reached middle age.

Brandling said] he would not interfere between Mr King and I. I said, "Good day, Jentelmen", and went to Heaton.

Ingaged at Backworth <77>

Joseph Smith told me they wanted a Waggon wright at Backworth [Colliery].* I set off [there] and was Ingaged and came and started at One a clock and I worked quite Savege. I would not Stop to Speek to no One [for] 10 days [? until] after I got paid and al Clear[ed] up. And got my tool Chist to Newcastle and from ther to Backworth. September the 13, 1818, Anthony Errington and [his] son, Anth Fenwick Errington, [set] to work at Backworth Colliery at Making waggons. After the pit got in, the railway was to Lay and Incline to make. I was in the pit 12 Months and more.

Backworth pit fire <78>

We had Lentheaned the plane [by] 2 piller[s] and [on] the Friday and Setorday [set] Double doores [i.e. air doors] above the Incline wheal. 3 Men went down at 12 Oclock to Clean the boards of stones. When they Opened the first door, it instently fired [= exploded] and Blood [= blowed] boath doors down and Burn[ed] the men. Before 1 Oclock I got to the pit. Roger Hutton, One of the Men that was burnt, was at the top of the pit. The Other Came to the botum and Came up. They boath Died. Mr Olliver Came to me and said, durst I go down with him? I said, yes. Mr Olivr and I worked near 2 hours before any help Came. We got the doors up and the Bretishes [= ventilation

---

* Backworth Colliery, in a previously unworked part of the coalfield six miles NW of Newcastle, was sunk in 1817-1818, and a waggonway of nearly five miles was constructed, to connect with an earlier, shorter waggonway to the Tyne. A.E. now introduces his second son to his trade, his first son having given it up <**56**>.

partitions] and Stopens [stoppings = ventilation walls] put up with deals.* The pit got to work again and I had to atend the railway.

A promise to drink no more <79>

The pay Night I got my Money, and One William Grame [only] Left off [drinking] at 4 Oclock and [was] meary [merry], [so] I took Care of him and Led him home. The hous had Stone Steps at the End and I got him into the house. The botum Step was brok[en] at the End of the handrail, and the fals step made me fall. And a Grames girl said, "Mother, the mens Drunk and fallen down". She brought a light, [but] I had Lost Nothing and I got home at 8 Clock. [The next day] My doughter Sarah had to go to the well for worter and Grames doughter [was] in Company.** When [the pot was] filled Sarah was to help her [i.e. Graham´s daughter] [to put it] on [her head] first and then She was to Lift with One hand to help Sarah On. [But] She walked away and Said She [i.e. Sarah] might go for her Drunken farther to help her on. [So] she had to wait One hour when a Women Came and helped the [pot] on and Came [home] with her. She was Crying and relating to Mother when I got home from work. This Praid upon my mind and I made my Daughter a promise I would Drink no more. I Vowed to God to not taste sperets, bear nor Small bear for One year. Which I kept for 13 month. And then at Warker [Colliery] I had imbibed gass in my Cheast and was bad for 3 weeks. And [to] get working, I had to go [to work] One Sunday night and 2 depeties [= foremen] was to go with me. They was at the publick house and had hot ale and Brandy. One said [to me], "Thee is Dissy on thy feet, drink that!". A glas ofered, I took it and had to run to the door to

---

* In order to disperse any remaining gas, it was essential after an explosion to restore the circulation of air by re-erecting the partitions and walls that directed the path of the air.

** Sarah, A.E.'s daughter by Jane Richardson, is now aged about 8.

throw up. [But] I got the Second and it stoped. I returned God thanks. As a Medison I had taken it. I had recourse to the Same when poorly and thus Ended my Vow.

Tinmouth Castle inchantment <80>

There was a Ladey, a profetess, who had pamphlets prented [saying] the Inchant[ment] had to be broke on the Tinmouth races and Faire day by a Dane.* Mr Talor of Backworth Aplied to the wastemen [of his colliery for men] to go and travell the Subteraneus pasages under the Castle.** No One would go, to go first. Mr Taylor wanted One [of them] to take the Charge and mark down the Lenths of Such place[s] [as were found]. I told him I durst go first. And [there were] 6 [men going] and I was 7, and a talor belongen North Shealds [went with us]. We prepared 2 Davee Lamps, 2 Steelmills, 2 dried hadecks, and 2 lb [of] Candles.*** And On the fair day I went to a Spice Shop to get the

---

* The castle at Tynemouth, the settlement on the north side of the mouth of the Tyne, was reconstructed in the seventeenth century on an earlier site, and was later a barracks. The nature of the "inchantment" is not explained in the text, and the pamphlet A.E. saw has not been traced. But earlier and contemporary references to an enchantment, while equally vague, refer specifically to the underground passageway known as 'Jingling Man's Hole' (Mackenzie 1825, 2:438; Richardson 1846, 7:390-5; Tomlinson 1888, pp.50-1; Shields Daily News, 27.8.1930, 30.8.1930; NSCL, Taggart typescript, p.32). This seems to be the second passageway A.E. explored. The Danes occupied the mouth of the Tyne in the ninth century and a tradition alleged that the castle originated as a Danish stronghold (Craster 1907, pp.40,154) - perhaps explaining the reference to "a Dane". Tynemouth fair was held in April and November.

** Thomas Taylor was the owner of Backworth Colliery.

*** A.E. lists the underground lighting equipment of the period. Candles were in general use. Where there was gas, steelmills, which threw out a shower of sparks and were considered (wrongly) to be incapable of igniting gas, were sometimes employed. As a last resort, miners occasionally tried to work by the phosphorescent glow thrown out by stale fish. The Davy lamp, the first safety lamp, had only been invented two or three years before this episode, and was not yet widely used.

[first] Davee Lighted and the woman seeing [it] Cried out it was an invisable [= ? invincible] lamp. On taken the second [lamp] she Observed my read [= red] hair and She Shouted, "There is the Dane going to break the Inchant!". We Decended a Shaft 4 feet Squair and to the depth of 12 feet, the botum being fill[ed] with Stones. There was an Archway for 12 or 16 yds, Creeping way, and then the hight was 7 feet, walled with huen Stone. From the Shaft to the End where the wall seased and the Soile had fallen in was 50 yds. We lookd [but could] Se no more there. There was One more hole on the Seebanks. We got all in, Creeping way, 20 yards, when we got into a place 16 feet Long and 12 broad with 5 Driftes [= passages] from this. Looking up I observed day Light through the brier bushes growing over the top [of] this. I then Percieved that it had bean used in the time of Seage as a Draw well. There was a Much decaid Sestron [= cistern] at the botum and there [were] Stone Spouts up Each Drift at the botum of a Seam of Coal and Segar clay. The warter was in the Clay and had Driped into the Sistron. On Looking round One of the Men Said there was a devel´s Imp. I said, "Catch him, [but], heare, take the taylor away first". On which he Clung to my arm. This was a Large Spidard, its Leggs before and behind covred 6 inches, the body being near One inch Long, quite Black. On searchen, we found thre[e] of [the] same. We had no plase to put them than a tobacco box. We Could se no more. We then thought to return, and going to the Tinmouth Inn, the Landlord made a through fare of us [??], and the Instruments up, One pair Stairs down, the Other up [??]. We had bear, Brandey, and rum, and Jin for the Ladies as they pased. And it was who was to be first with hauf Crowns on the table. Each and Evry one [was] served as they pased. [It was said that] the whole [adventure] would be advertised [in the newspapers] on the Setorday first [following]. The Ladies, onn seeing the Large Spiders, fear was on some of them, [but] I told them not to be feared, they was only spidrds. We paid the Land Lord and had 18s to good.

Wee agreed to go to the Wormouth Bridge and have Super which we did privet, the taylor with us, then we returned home.*

The Joues buriel ground <81>

At Backworth I got a Line from My Brother William to go and have diner [? with him] on board the Elexander of London Loading Coals at Cocks Lodge spout. I aprised the master [whom] he was to set on in my roome [= place]. I got diner [on the ship] and then went [? on the ship] to Shealds. [The ship] did go to Se that night. [There was] no moon and I stoped with him [i.e. brother William] and Sisters to ½ past 8 PM.** [To go home,] I had to go to Chirton and up the Billimill lane. It was dark but the road was Drie. I Could just decern the house tops and tree tops. The Jues has a buoriell plase [there].*** And on Coming near [it], I had the Contemplation [that] they would bee up on the Day of resorection as well as those in

---

* The underground passages have been explored several times over the centuries but records are exiguous. An 1849 guide book spoke of an exploration "not many years ago" (Gibson 1849, p.130); a local writer c.1930 described, from an unnamed diary, an exploration in 1778, and also another, allegedly in 1837 (NSCL, Taggart typescript, pp.30-32). All three accounts bear some resemblance to A.E.'s in that they mention arched ways, a well, and a chamber or chambers; but it is possible that the later references arc in fact to A.E.'s expedition (although not based on his account).

** The Coxlodge 'spout' or staithe was located at Wallsend. It appears that William Errington's employment was in the coal shipping trade, and that he lived at North Shields. In 1834 a William Errington of North Shields was a shipmaster (Directory 1834). Unfortunately this passage of the manuscript is garbled, but it seems that the ship, once loaded, passed down-river and landed A.E. at North Shields. If the ship went to sea but William stayed at North Shields, he cannot have been the captain, as otherwise might appear to have been the case.

*** A.E.'s five-mile walk back to Backworth from North Shields passed through Chirton and Billy Mill. A Jews' burial ground existed in Billy Mill Lane until the 1850s (Mackenzie 1825, 2:457; Craster 1907, p.378).

the Church yard. When [suddenly] there was as a ball of Fire Come from the west Crosing 5 yd before me, and I felt the heat [as] it Daselled my Eyes. I kept walking forward when I found I was on the back of Somthing. At this my hare Lifted my hat. It was runing away with me. It then Started to Bray and [I realised] it was an ass. I got off and said, "Poor sillie thing, thee has given me a flay [= fright]!" I got very well home and went to work. At 2 Clock Mr Oliver came. I told him I had brought a Couple of beskits and a Slice of Corned Beef, and I related to him my Jurnie and return. Mr Buddle came [at] 4 Clock, but I did not Se him, [so] he wated to 10 oclock to have the Storey from my own mouth. Mr T Taylor and he got a good Larf at the Cudey [= ass] and I.

Under my vow <82>

At this time I was under my Vow and the pit holed to a bore hole and the warter was beating the Ingine. Al hand[s] was Called to get prepared for pluging the hole, and all was redey for Drawing away the horses [from the seam]. [But] The water rather abated, and after 4 hours 12 men got the hole plugged on Time and all [getting] as wet as warter Could make us. We was [told] to go and get brandy and Ingoy our selves. [But] I got a pint of new Milk and Stoped a short time and returned home.

To Worker Colliery as a more faithful servent <83>

Mr Thos King was Viewer at Worker [Colliery, where] he got 2 Ingines and 2 pits Opened to the botum.* [But] he gave the Owners up and they had to aplie to John

* Walker Colliery, an old mine, was located three miles east of Newcastle, on the north side of the river, almost opposite Felling. A.E. has told us that his health was affected by gas while at Walker <79>.

Straker to be Viewer, which he accepted. One John Foster and [his] 2 Suns had Laid the railway down, and had 20 waggons to make when Mr Straker Entered. [But] he [i.e. Foster] had drawn as much money as [if] the whole [job] had bean finished. [So] he [i.e. Mr Straker] Sent for me, and I was Ingaged as a more faithful servent, Mr Taylor granting [me leave] to go, as he was Senier Master.

fidle [illegible] at [illegible] the Men had been at work from 8 Clock P.M. on Sonday
and at 6 Clock monday Morning went to headon pablick to have Bref Stakes
and Bread and the meary Clock thein was a Scotch Lad Comin booth
fast and his Mother they had been 13 days in Newcastle and had poor
Luck he had the fiddle I asked the Mother had they had any brickfast
no nor not One peney in the pocket [illegible] with Compasion for the
poor and [illegible] I got a plate and gathered 14 Shilings for them the back
[illegible] and returned thanks to God for his mercies to them

I Saw them Once after and 4 years after I Saw the Lad [illegible]
mother was [illegible] he had a Confinement [illegible] he [illegible]
he Suported and I never Saw him more

FROM THE ERRINGTON MANUSCRIPT (unnumbered page, anecdote 64)

L I S T O F A N E C D O T E S

1. The reason of my wrighting
2. My father and my mother
3. Anthony Errington born
4. Put to skool
5. Burried alive
6. An experiment
7. Singler case of birdnesting
8. A pistle found
9. A good doctriss
10. Learning to swim
11. Saving a man at a pit
12. Saving men on the waggonway
13. To be a blacksmith
14. The pit fires
15. To be a farmer
16. Condemned to walk the plank
17. Apprentice to my father
18. Working with the adge
19. Singlar profisy
20. The Power of Christ reineth
21. The seccond case
22. Third case
23. Brotherly love
24. A stolen gallaway
25. The tayler's goose
26. Singlar meeting
27. Manchester waggon and eight horses and driver - and poney
28. Singlar escape
29. Robert Burns the Scotch bard
30. Sunrise in the morning - a woman stopped
31. To change of life
32. Hollihill pit blower
33. Hanging in the shaft
34. Hillpit fire and loss of life
35. Grand state question
36. Felling Colliery 1803 - the bone setter
37. Bad money
38. Newcastle date
39. Proverbe
40. Pill doctor
41. Assaulted at Barkers
42. The Penypie jack in Newcastle
43. Holing the waste
44. Bad company
45. Singlar battle
46. Death of wife
47. A mason saved
48. Divine providence
49. Recalled from Benwell
50. John Pit railway - April the 18 1811
51. Hardware shop in The Side
52. Temptation to bee reserection man
53. A man prevented from hanging
54. George Hawkes drunk
55. Shifting to Percy Main
56. My son very ill
57. 1813 Howden pit blasted
58. Society esteblished at Percy Main for the betar ventalation of the mine
59. Saved by a dog
60. The dog stolen
61. Fire in Howden pit
62. Bodily fear of six men
63. A Scotch piper
64. A Scotch fiddler
65. Out from the teapot
66. Intrige of a woman
67. No more peace at Percy Main

LIST OF ANECDOTES

Brunton
Brunton
North
Gosforth
Killingworth Moor Inclosed
Shire Moor
Walbottle Moor
Newbigin
Brunton
Fawdon
Cocksload
Longbenton
105 Fas
45 Fas
College Lands
Mr Lake
Mr Wilson
South
Gosforth
Longbenton
Kenton
Willington
121 Fas
High
Flatworth
Piggs
Roman
Walls End
Montague Main
60 Fas
Fenham
Race Ground
The Town Moor
Jesmond
Biggs Main
90 Fas
Walls End
Flatworth
Heaton Main
72 Fas
Nr Elswick
Shield Field
95 Fas
Adairs Main
Byker
Walker
100 Fas
Hebburn
Bladon Main
Elswick
St Anthons
Swalwell
65 Fas
Tyne Main
Gateshead
Monkton
Whickham
Red Heugh
Park
Brandling Main
Heworth
Newton
Garth
Hedworth
Inclosed
Field House
Township
Wardley
Lands
Saltwell Side
Whickham
Common
Low Moor
60 Fas
Gateshead
Fell
Mr Robson
White House
West Boldon
Sheriff Hill

ANALYTICAL COMMENTARIES

LIST OF COMMENTARIES

Introduction

1. Table of chronology
2. Topography
3. Technology of contemporary Tyneside coalmining
4. A.E.'s family, homes, and domestic arrangements
5. A.E.'s religion and moral values
6. A.E.'s education and training
7. A.E.'s conditions of work and earnings
8. Food and drink
9. Health
10. Crime
11. Anthony Errington - the man and his writing
12. The manuscript and its editing

A N A L Y T I C A L   C O M M E N T A R I E S

INTRODUCTION

The Commentaries that follow represent, in the first instance, **extended subject indexes.** Under each head, all the references within the autobiography to that theme are cited and linked, and a summary of the theme is attempted. References are made to individual anecdotes by number; quotations from anecdotes are given in simplified form, without editorial additions being signalled. Inevitably the material overlaps between the various commentaries and hence there is some repetition.

But the larger Commentaries also include **explanatory information,** supplementary to that in the footnotes, and based to some extent on wider sources, including documentation other than the manuscript text. Occasionally I have thought it proper to invoke the broader historical context, but since so many of the received generalised interpretations of the social history of the period are based on inadequate research and selective evidence, this has been done sparingly.

The Commentaries conclude with a description of the manuscript and a statement of the editing procedures.

Some readers may be surprised at the amount of information that can be recovered from the extant records regarding the life of an individual who existed two centuries ago, who occupied a lowly station in society, and who until now was unknown to posterity. However, I must make it clear that the research undertaken for this edition has only scratched the surface of the extant records, hence that the edition leaves many loose ends, all of which could be, and should be, noted and then investigated by other historians, and probably by local historians in particular. More extended research in the Tyneside newspapers, in local government records, and in Roman Catholic and Anglican church registers would almost surely add to the documentation of the Errington family - not to mention those sources which I have not had convenient access to or have entirely failed to realise were relevant. Tyneside has a particularly rich collection of records of the coalmining industry, and these include the work records of John Buddle, A.E.'s acquaintance. In 1950-1951, when the papers of Buddle and other viewers of the period, in several hundred volumes, were held at Neville Hall, Newcastle, I was permitted to examine and use them, and I prepared a checklist; but I have not had the opportunity latterly to re-inspect more than a tiny corner of this material, despite the recent transfer of the papers to the Northumberland County Record Office where they are more accessible to local historians. These papers contain extensive material on Felling Colliery and on all the other collieries at which A.E. worked, and if they were combed through, references to our author as waggonway wright and to the underground episodes he describes would almost certainly be found.

I commend these tasks to the eager band of historical researchers nowadays to be found in local history societies, family and population history societies, industrial and labour history societies. The present edition is printed - as will be obvious to the reader - from camera-ready copy prepared by myself. Whatever the disadvantages of this procedure, it has the advantage that the edition can, if necessary, be revised and re-issued with fair ease. Corrections and additions from readers and researchers would therefore be welcome.

Having sat on A.E.'s autobiog phy for thirty and more years, I see it into print with an apolc ' for the limitations of the research. But the need to draw a line across the page - eventually - strikes every historian. The anxious words of a past historian of Newcastle have some relevance. "Topographical history is peculiarly exposed to general and severe criticism; for though it is scarcely possible to obtain unerring accuracy in the numerous and diversified notices which compose such work, yet no error, however trifling, can escape detection" [Mackenzie 1827, Preface].

May it be so !

COMMENTARIES

A SAMPLE OF THE TEXT IN UNALTERED TRANSCRIPTION FROM THE MANUSCRIPT
(see the reproduction on p.115)

I dwelled at howden 16 Men had bean at work from 1 Clock P M on Sonday

and at 6 Oclock monday Morning went to howden publick to have Beef Stakes

and Bread and the near 7 Clock their was a Scotch Lad Lam in boath

feet and his Mother they had bean 13 days in Newcastle and had poor

Luck he had the fiddle I asked the Mother had they any brickfast

no nor not One peney in the pocket tuched with Compasion for the

poor and Nedey I got a plate and gethered 14 Shilings for them the boath

nealed down and returned thanks to god for his mercies to them

I saw them Once after and 7 years after Saw the Lad his

mother was dead he had a Orfin girl 8 or 9 yeares of Age which

he Suported and I never Saw him more

(anecdote **64**)

TABLE OF CHRONOLOGY

(A.E. is the subject unless otherwise stated.)

1732	father born <2>
1754-1799	father works as waggonway-wright <**2**>
1766	father marries at Heworth <2>
1766-1818	father's home at Felling until death <**2**>
1777	Felling colliery begun <2>
1778	Anthony, fourth and last son born <**3**>
c.1785	at school <**4-10**>
1787	school-mistress dies <**9**>
1791	aged 13, tries blacksmithing and farming, and assists father down the pit <**13-15**>
1792	apprenticed to father <17>
1792-1809	works at **Felling Colliery**
1793	brother George's regiment passes <**23**>
? 1795	meets a man who claims to be Robert Burns <**29**>
1798	marries Ann Hindmarsh <**31**>
1798-1806	four children born
1799	father lays rope-walk waggonway <**2**>
1799	father retires from pit work, takes small farm <2>
? 1801	wins newspaper competition <**35**>
1802	mother dies
1803	Hollyhill pit explosion <**34**>
1803	child saved <**36**>
1807	Discovery pit inundated <**43**>
1809	wife dies <**46**>
1809	transferred for short period to **Benwell Colliery** <46>
1809	saves falling man <**47**>
? 1809/1810	returns to **Felling Colliery** <49>
1810	child by Jane Richardson
1810	celebration of waggonway from John pit, Felling <**50**>
1811	saves would-be suicide <**53**>
1811	leaves Felling colliery, joins **Percy Main Colliery**, with son Robert <**55**>

1812	Robert ill, apprenticed to painter <**56**>
1813	averts explosion at Howdon pit <**57**>
1813	makes proposals for mines safety to new society <**58**>
1814	leaves Percy Main Colliery, joins
	Fawdon Colliery <**67-68**>
1814	daughter Ann dies <**68**>
1815	sees vision predicting Heaton pit disaster <**70**>
? 1816	leaves Fawdon Colliery, joins
	Gosforth Colliery <**74**>
1817	purchase of Coxlodge Colliery celebrated <**74**>
1818	leaves Gosforth/Coxlodge Colliery, joins
	Backworth Colliery, with second son <**76-77**>
1818	father dies, aged 86 <**75**>
? 1818	brother George leaves army <**23**>
1818	marries Mary Pearson <**75**>
1819-1831	four children born
1819	child baptised at North Shields
? c.1820	Tynemouth Castle exploration <**80**>
1821	child baptised at Newcastle
?	leaves Backworth Colliery, joins
	Walker Colliery <**83**>
1825	child baptised at North Shields
1827	milkmaid met 13 years after 1814 <**69**>
1831	child baptised at Newcastle
1834	brother George dies <**23**>
1843	Buddle dies
1848	A.E. dies

TOPOGRAPHY

Tyneside

A.E. was born at Felling, near Gateshead, on Tyneside. Up to his mid forties, when his autobiography ends, according to that record he had never travelled more than twenty miles away from his birthplace, with one possible exception. As a child, he attended school in High Felling and at neighbouring Low Heworth <**4,5,10**>, and he reports having visited in this locality Friars Goose Wood <**4**>, Heworth Church <**5,9**>, and Heworth Burn near the Sunderland Road bridge <**7**>. A little further afield, he went down-river twice, once to South Shields, apparently on foot, where he bathed at the "Yow Hole", the second time to North Shields in a boat <**10,16**>; and he was sent with a message to Sheriff Hill Colliery staithe, which was at Redheugh, on the west side of Gateshead <**12**>. In his early teens, he made four reported journeys through Gateshead and across the Tyne Bridge to Newcastle, once with his soldier brother, once on an errand for his father, once to meet his parents who were probably attending a fair, and once to go to church on a Sunday <**23,25,27,26**> - it is likely that his church visits to Newcastle were regular. However, in the same period of life he was sent away from home to try out two jobs, the first as a blacksmith in "the Cuntry" at "Bublok", an unidentified place probably in north Durham (but if further afield possibly providing an exception to his limited range of travel), the second as a "farmer", that is, a farm labourer, at Hylton Castle, between Felling and Sunderland <**15,13**>. A little later, probably when he was about fifteen, he made his way to Sunderland, about eight miles east of Felling, to see his soldier brother at the barracks there <**23**>.

In adult life A.E. continued to live at Felling - from 1798 in a family home of his own - up to the time of his wife's death in 1809. Thereafter, with a short break in 1810-1811 when he returned to Felling, A.E. worked at collieries on the north side of the Tyne - Benwell, Percy Main, Fawdon, Gosforth/Coxlodge, Backworth, and Walker. At least at first he moved rapidly, between 1811 and 1818 working at Percy Main, Fawdon and Gosforth/Coxlodge. His lengths of service at Backworth and Walker are not recorded in the autobiography, and it is uncertain whether he moved from Backworth before 1822 or after 1825 (he had children baptised at Newcastle in 1821 and at North Shields in 1825). Walker and Percy Main lay down-river from Newcastle, the other collieries lay further back from the river, Fawdon, Gosforth/Coxlodge and Backworth being several miles NW, NNW and NE of Newcastle as it then was. But at no time was A.E. out of walking distance of the centre of Newcastle. When living north of the Tyne, from 1811 perhaps up to his father's death in 1818, he returned to Felling at weekends to see his children and thus regularly passed through Newcastle. But he also regularly visited it for shopping and convivial purposes. It appears that he walked the distances, which were up to seven miles. He specifies walking between Gateshead and Felling <44>, South Shields and Felling <53>, Fawdon and Newcastle <**70,71**>, and North Shields and Backworth <**81**>, and he never refers to travelling in a cart or carriage. But he does mention going up-river, from Howdon to Newcastle, in a wherry, <**63**>, and to crossing the river in a wherry from Newcastle to Hebburn, on an occasion when his further journey to Percy Main was almost certainly also by boat <**65**>; and it may be assumed that, when working on the north side, he often crossed to Felling by the wherries which provided a regular ferry service [Mackenzie 1827, p.722]. He further mentions travelling up-river, in an emergency, by steam-boat (allegedly in 1812, a misrecollection, since there were no steam-boats on the river before 1814), probably from North Shields to Newcastle <**56**>. He once boarded a coal vessel at Coxlodge Colliery staithe, near Wallsend, and apparently travelled on the ship down-river to North Shields, where his brother William lived <**59**>.

Thus A.E. was acquainted with all the localities of the lower Tyne, from Newcastle and Gateshead down to the river mouth, at Tynemouth Bar <**10**> and South Shields. On the south bank and travelling from Felling, he twice records visiting South Shields - once specifically on colliery business. On the second occasion he went by way of a street called "the Sleck Row" <**53,69**>, which was near the head of Jarrow Slake where now are found Slake Road (Jarrow) and Slake Terrace (South Shields) (at one time a 'Slake Row' was joined to Turncoat Row in South Shields [Directory 1865]). On the north bank and travelling from Percy Main or Backworth, he records visiting

North Shields several times (via Chirton and/or Billy Mill) <**59,66,81**> and Tynemouth once <**80**>. As a child he had gathered mussels at the Low Lights at North Shields <**16**>, perhaps on the rocks called Mussel Scarp. As an adult, only in one instance does he record working outside the lower Tyne district, and then just outside: in 1811 he built a railway (not a colliery line) at Southwick, five miles west of Felling and near Sunderland <**46**>.

Newcastle

The Newcastle which is the scene of many of A.E.'s adventures was not the Grainger/Dobson showpiece so much admired by Pevsner in the 1950s and still largely in existence, if beleaguered. But nor was it the medieval walled town which survived until the eighteenth century and was recorded by Brand in 1789. Writing in the 1820s, the historian, Eneas Mackenzie, celebrated the "improvements" of the last half-century - that is, in A.E.'s lifetime - including the demolition of a large part of the town walls and all the gates. The Tyne Bridge joining Newcastle and Gateshead, over which A.E. passed so frequently, was entirely rebuilt in the 1770s and widened in 1801. (But there appears to be no historical reference other than A.E.'s to the public privy at the end of the bridge <**38**>). The West Gate, where A.E. saw off his soldier brother in the early 1790s, was pulled down in 1811: the Half Moon Battery, where A.E. had his first Newcastle adventure c.1790 <**25**>, was soon afterwards demolished [Mackenzie 1827, pp.109,168]. The Old Flesh Market, an open-air site for butchers' stalls, noted by A.E. as the location of the 'Dog and Duck' where the pseudo-Burns was encountered in the 1790s <**29**>, was gradually put out of use by the New Flesh Market, which had built-in shops, opened in 1808. The two markets were linked by Drury Lane, perhaps A.E.'s "Dowrie Lane", where he and his wife had a meal in an eating-house, after buying meat at the "Bucher Market" <**45**>. On another occasion, A.E. met a couple in the "Bucher Market" <**54**> - it is not clear in either episode which market was meant [Mackenzie 1827, p.176; Middlebrook 1950, p.151]. An older improvement, the Infirmary on Forth Banks, built in the 1750s, was walked past by A.E. on his childhood Sunday adventure <**26**>.

However, many of A.E.'s references are to traditional features of the town, particularly those near the river - features which survived his life-time, and which, indeed, are still mostly represented by presentday features, albeit often in attenuated form. The Old Customs House in Sandhill, replaced in 1765 by the New Customs House on the Quay [Mackenzie 1827, p.720], was where A.E. met the pieman <42>. On his Sunday adventure, walking from the bridge A.E. passed through Sandhill, an open space, and along The Close <**26**>: he heard the bells of St Nicholas' church, and in later life he almost witnessed body-snatching in St Nicholas' churchyard <**52**>. He visited public houses on the Quay <**63,65**>; and a public house and a hardware shop in The Side <**51,52**>. His brother-in-law lived in High Friar Street <35>.

But once A.E. had moved to collieries north of Newcastle, his anecdotes also mention localities on the northern outskirts of the town. At Barras Bridge in 1818, A.E., who was apparently then living at Gosforth and was on his way to the town centre on a free Saturday, met a rival travelling with his market purchases in the other direction, to "Dean Houses" <**76**>, which stood above a quarry in Jesmond Dene [Mackenzie 1825, 2:470]. Barras Bridge was to be rebuilt and widened in the following year [Mackenzie 1827, p.192] - today the non-existent 'bridge' is a mere run-up to a busy road junction. The Town Moor, then on the northern outskirts, was the scene of visions <**29,70,71**> and also of the Cow Hill Fair <**72**>; as A.E. mentions the grandstand on the Race-Course <**71**>, which was built in 1800 and served as a tavern during the summer season - it stood on the north edge of the moor, near where Kenton Road now turns off Grandstand Road [Middlebrook 1950, p.151]. The first vision was seen on "the foot path from the Kowgate along the West end of the Moor" when A.E. was walking from Newcastle "home to Kenton"; the second vision was seen when again walking home - specifically before A.E. reached the Cowgate and after "Crosing to the grandstand" <**70,71**>. Cowgate, formerly the West Cow Gate of the Nuns' Moor, was by A.E.'s time a junction on the Ponteland turnpike road, and A.E.'s route between Cowgate and the town centre could have been by this road and via Gallowgate, and not by Barras Bridge [Mackenzie 1827, p. 710]. Whichever way he came, A.E.'s telescoped itinerary ("set off" from central Newcastle by "the foot path from the Kowgate" on the outskirts) most probably means that the first vision was seen north of the Cowgate, hence the reference to the "North hedge". The second vision was seen "Coming up the North side" of the moor as A.E. was cutting across the moor in a northward direction, so that he at first thought it a burial party from Gosforth still further north. Since both parties of apparitions were moving west, and both were seen on the same north side of the moor, although some distance apart, they were following approximately the same track, which no doubt further impressed A.E.

Further out than the contemporary outskirts of Newcastle, to the north of the town A.E. mentions Benton Moor <**1**>, Jesmond <**41,70**>, Gosforth and Little Gosforth <**29,74,76**>, Fawdon <67,68,71-4>, Kenton <**70**> (including "Hensharber" <**68**>), Coxlodge <**74,75**>, Longbenton <**76**>, and Backworth <**77,78,81**> - all except the last now suburbs of Newcastle. To the west of Newcastle, he mentions Benwell and Fenham <**48**>, and to the east Heaton <**70,71,76**> - all also now suburbs. The episodes involving these outer localities are generally related to A.E.'s work at various collieries.

While working at Benwell in 1809-1810 he lived at Fenham Lodge, perhaps the lodge of Fenham Hall <**48**>. While at Percy Main Colliery in 1811-1814, he lived, at least for a time, at Howdon, in Howdon Row <**59,63**>. At Fawdon Colliery in 1814-(?)1817, he lived at "Hensharber" (Hens Harbour), a row of houses (shown on the 1864 OS six-inch map) at the east end of Kenton, the village before Fawdon, <**68,70**>. In 1825 Kenton was described as "a large pit village, containing 134 dwelling houses", whereas "High Fawdon, or Fawdon Square, consists of about 70 pit houses, and Low Fawdon of about 20 houses" [Mackenzie 1825, 2:472-3]. Both Kenton and Fawdon then lay within the parish of Gosforth, hence A.E. states that his daughter was buried at Gosforth <**68**> - but this was perhaps a misrecollection, since the burial cannot be traced in the Gosforth parish register. A.E. later worked at Backworth and Walker collieries and presumably lived near each, but he specifies no exact location.

Gateshead

"The ancient borough of Gateshead consists chiefly of one long street, descending to the bridge leading over Tyne to Newcastle"

[Hutchinson 1787, 2:373]. Even in the 1820s this description of the 1780s required only limited modification. From Felling, A.E. travelled a long mile into Gateshead along the Sunderland road <**7,27**>, a turnpike after 1796 [Bailey 1810, p.270]. This joined the southern and upper end of Gateshead's main street, the High Street (or simply Gateshead Street <54>), whose northern and lower end, named Bottle Bank, ran down to the bridge. Making his way down the High Street one day, and approaching the cross-street still named Jackson Street, A.E. saw a road waggon about to collapse <**27**>. Four months later, A.E. was again in Gateshead, this time with his father: probably father and son visited Gateshead regularly, for business reasons, since A.E. knew where the nearest blacksmith's shop was situated <**27**>. Later in his life, A.E. twice meets individuals in Gateshead, but the autobiography does not specify the circumstances or exact locality <**53,54**>. Of the three public houses at Gateshead mentioned by A.E., all were situated in the lower part of the main street, one giving its name to Black Bull Yard and a second to Half Moon Street <**14,47**>. The Freemasons' Lodge at Gatehead, meeting probably in the Blue Bell, Bridge Street (it met there in 1794 but in 1811 at the Goat, Bottle Bank) was visited at least once by A.E.: on this occasion, instead of returning to Felling by the Sunderland Road, A.E. took a short cut nearer the shore-line and past Friars Goose staithe, apparently along a footpath <**44**>. This path must have gone near Friars Goose Wood which A.E. had visited as a child, and which at that date had ploughed fields around it <**4**>. As a child, he had also walked down the Sheriff Hill Colliery waggonway, which, as he says, at this date went past the Windmill Hills, west of the High Street, on its way to the river; but where exactly along this route the accident that he reports occurred is not clear <**12**>. Also as a child, he had worked with his father on the South Shore - the Gateshead shore east of Tyne Bridge - at Chapman's ropery <**2,18**>. When he came to be apprenticed to his father, A.E.'s bondsman was a tailor of Gateshead Fell <**17**>.

The ironworks of Hawks and Co. were the main industrial enterprise in Gateshead at the beginning of the nineteenth century. As a child, A.E. met at Gateshead "9 of Mr Haukses Smiths" <27>; in 1809 he had a contact there with an "Edge tool maker for the Hawks" and some of "Hawkes Smiths" <47>; and at an uncertain date he met in Newcastle "2 of Squire <?> Hawks Blacksmiths" and supported one of them in a fight. For this, A.E. was thanked by "Mr Hawks" in person when four months later he "had to order some Waggon wheals at Mr Hawkses", presumably at Gateshead <**45**>. Although the Erringtons thus made use of Gateshead heavy industry, the autobiography does not specifically mention any other purchases in the town - we find A.E. shopping in Newcastle for meat and even for small hardware. The drunken George Hawkes and his wife, apparently residents of Heworth, who owned a cart in which they travelled, also bought their meat in Newcastle <**54**>. However, a reference in this anecdote to A.E.'s aged father and his shopping does not make it clear whether the old man now shopped in Newcastle or Gateshead <54>.

Felling

On its way east from Gateshead to Felling, the Sunderland turnpike road passed "the Robers Corner" <**54**>, Robbers Corner being the point on the road where the entrance lodge of Park House then stood [Manders 1975, p.130]. Around 1790 the Sunderland road was one of those "infested at night by foot pads" <**8**> - but at this date the reference may have been to the old Sunderland road which on higher ground came along Split Crow Lane and passed through High Felling to Heworth Grange. At one stage in his working life, A.E.'s father "made the waggons for the Old Faud Colliery, near Gateshead" <**2**>: Old Fold is now a district linking Gateshead and Felling. A.E. courted his wife at Felling Gate <**31**>: this was almost certainly not a railway gate, but Kirton's Toll Gate on the turnpike - the stated residence of the Mr.Sill for whom Ann Hindmarsh was housekeeper, at least when he died a few years later [Heworth burial register, 6.2.1801].

When first married, A.E. lived at High Felling, but in 1799 moved to Low Felling, perhaps immediately to "a Cotage" there <**31,44**>. Low Felling, the residential centre of the three Fellings, was described in the 1830s, shortly after the autobiography concludes, as follows: "It contains a few new houses, and many cottages for the collieries, which, with the small gardens attached, give an aspect of comfort and industry to the place. Here are three surgeons, one victualler, several mechanics, and a poor-house" [Mackenzie 1834, 1:22].

The location of A.E.'s cottage at Low Felling can be worked out. The John pit at Felling, "near the Sunderland road" <**2**> - more precisely, "on the north side of the Sunderland road, and half way between Felling Tollbar and Felling Hall" [Hodgson 1813, p.6] - had a railway which ran 951 yds. to Paddock Hall turn <**50**>. As shown on 1826 and c.1840 plans [GCL, Bell plans] and the 1856 OS map, the John pit waggonway had a slight bend at approximately the stated distance from the pit, and this must have been Paddock Hall turn. The name 'Paddock Hall' is not remembered locally but appears in an 1801 entry in the parish register [Heworth burial register, 9.1.1801]. A 'paddock' being a frog, there may have been a connection with ponds shown near the turn on the 1826 plan and with the name 'The Baths' shown on the 1856 map and surviving in the name of Bath Road. The fact that the railway was built only to Paddock Hall turn suggests that it joined up with an earlier line, and an 1806 document plans a waggonway from a shaft apparently being sunk and probably the later-named John pit "to where it will join the Ann pit way 400 yds." [NCRO, Watson Papers 3/31/106]. At an unstated date and possibly before the opening of the John pit stretch of waggonway, A.E. rescued a man from a "railway guter" and took him to Paddock Hall and put him to bed, presumably in his own house <**40**>. This suggests that Paddock Hall was the locality of A.E.'s cottage. On the c.1840 plan the turn is shown near the site of the factory of Lee and Pattinson, and A.E. states that his cottage was where "Lees factory is at preasent" <**44**>. The chemical factory of Lee and Pattinson, established in 1833, and the ruins still known in the 1880s as 'Lees factory', was situated just west of what used to be the middle stretch of Brewery Lane before this road was closed off by recent industrial development [Campbell 1962, p.21; GCL, plans of the factory c.1840 and Taylor typescript, vol.2, p.119]. On the 1826 plan, the site of the future factory is shown as a piece of waste land, with a single building on the east side. The key to the plan described the site as "Pit Waste, and Cottage and Garden", and since no other dwelling was shown anywhere near the turn, presumably this was A.E.'s cottage. (If A.E.'s cottage did indeed have a garden, he never mentions it.)

As a child, A.E. "got aquanted with the Boys of Felling Shore" and this produced an adventure down-river <**16**>. He also worked with his father at Felling Shore, and from there was sent a mile or so to a farm at Carr Hill, above Felling, and to Squire Russell's house at Low Heworth (the Russells lived at Heworth Hall, "upon a commanding eminence" [Mackenzie 1834, p.17] but now overlooking a busy road), to help with butter-making <**20-22**>. A Felling Shore ship builder tried in vain to recruit him as an apprentice <**17**>. In the 1790s Felling Shore consisted of a small number of industrial establishments: forty years later it was described as "one of those populous manufacturing, and trading villages, which will probably soon form one continuous line along the bank of the Tyne from Gateshead to Bill Quay" [Mackenzie 1834, 1:24] - a correct prophecy.

One of A.E.'s Felling adventures happened when "going to diner along the Summerhouse row" <**36**> - this street was presumably named after the 'Summerhouse' of Heworth Hall, in Heworth Lane, and hence located on the Heworth side of Felling. It is therefore more likely that A.E. was going out to dinner than that he was returning home to dinner at the Low Felling cottage, since all the Felling pits were situated west of Heworth Lane. The location of the John pit has been stated above - its upcast shaft, the William pit, which played an essential part in the rescue after the 1812 disaster, is never mentioned by A.E. At the celebration of the opening of the John pit railway in 1810 an incident occurred which led to Mr Straker, the viewer, sending for a violin to Felling Hall <**50**>. Felling Hall, the Brandlings' chief residence until the 1760s, was given as the address of the previous Felling viewer, Thomas Dodds, in 1799, and as Straker's address in 1802 and 1809 [Heworth baptism register; NCRO, Watson Papers, 3/31/15]. The house was affected by mining subsidence [Mackenzie 1834, 1:22], and eventually pulled down, but is today commemorated in 'Felling Hall Gardens'. The exact location of some of the other Felling pits mentioned by A.E. is known. Thus, a shaft of the Hollyhill pit can still be seen, near the railway line, on the north side, in an angle between the main line and a former side line to a brickworks, nearly opposite Felling Park. (When A.E. built a railway of only 345 yds. from this pit, in order to carry coals to the river <**49**>, it must have joined another line, since the distance from this shaft to the river is over half a mile - it probably ran to the John pit, along the line of the present railway). The Keppel pit was located in High Felling, near where Crow Hall stood, and a health centre has been built around but not over the shaft. The Discovery pit was located "above High Felling" <30>, 500 yards SW of Felling Square (near where Victoria Avenue meets Nursery Lane); and the Venture pit was located towards Felling Shore. The locations of the Ann and Rodney pits are uncertain.

In 1799, A.E.'s father retired to "a small farm at Felling" <2>, and since A.E.'s mother died at Low Felling [Heworth burial register], the farm must have been there. However, when Robert Errington married in 1766, he and his wife "set up the first house" at High Felling, and "their Dwelled and Brought up 4 Suns and 5 Doughters" <2>. This ambiguous statement may mean that A.E.'s parents stayed in the same house at High Felling until his father retired, although that degree of immobility in a colliery community and for a man with a growing family would seem unlikely; or, alternatively, that they continued to live at High Felling but not always in the "first house". A.E. locates his own birth simply at Felling <2,3>. As a child of ten or eleven, A.E. reached the Hollihill pit on his way home from school at Low Heworth <6>. But, given schoolboy meanderings, this fails to inform us where in High Felling his home lay. However, in 1793, brother George's regiment, marching from Sunderland to Newcastle "had to pas the house" and some officers were invited into "the cotage" <23>. This probably indicates that the regiment was taking the old Sunderland road (i.e. Split Crow Lane) to Gateshead, despite its inclines, by coming up either High Felling Road (modern Felling High Street) or, perhaps more likely, over Holly Hill. If so, then the cottage in which the Erringtons lived was probably in the nucleus of High Felling village and bordering the old Sunderland road (a generation later the houses of this area were shown in detail on an 1826 plan [GCL, Bell plan 61]). High Felling was a very small hamlet in A.E.'s childhood, but later grew. "High Felling has recently been extended as almost to join Low Felling. Here is a Methodist chapel, two public houses, and butchers' and grocers' shops" [Mackenzie 1834, 1:24]. A.E. mentions the Bathe Well, in "Felling Dean, south side of the Sunderland road <7>, by which in this case he means the new Sunderland road. As shown on an 1843 plan and the 1856 OS map, two wells, now covered over, once existed off the south side of High Street, Felling, where a rivulet (or "Open Kennel") ran through a shallow valley, and the Bathe Well was located in front of modern Caxton House [GCL, Bell plan 76].

Just across the street is the site of the house where A.E.'s widow was living in 1851. If we may assume that this is where A.E. died, then he died within 200 yards of his most probable birth-place.

Although A.E. apparently gave up his house in Felling in either 1809 or 1810, his children continued to live in the village with their grandfather, presumably at the Low Felling "small farm", the younger ones for at least five or six years and perhaps until the old man died in 1818; and therefore A.E. visited Felling regularly during this period, as explicit or implicit references indicate <**50,52-4,56,60,68,71,75**>. It is not known when he returned to live at Felling, but it may have been by 1827 and he certainly seems to have been there in 1831. Possibly he stayed there until he retired from being a waggonway wright. What we do know is that he died at High Felling in 1848. His parents had also died at Felling, his mother after living there at least thirty-five years, his father after living there at least fifty years. His first wife died there, his second wife was born there and may well have died there. A.E.'s employers, the Brandlings, are still much commemorated at Felling, in half a dozen street names, the Brandling Hall Community Centre, and an architectural reference to the Brandling Railway in the old (now abandoned) railway station. Perhaps A.E. will also now come to be commemorated at Felling.

Felling was "in the Cheplry of Hueth in the parish of Jarrah" <**2**>: the same spelling, 'Hueth', was used in the contemporary register of the Catholic chapel at Newcastle. Apart from his childhood adventures at Heworth, we find A.E. escorting a drunken man there, and then drinking with him, presumably at a public house <**54**>; and again drinking at Low Heworth with a "wise man" <**53**>. A.E. was buried in Heworth churchyard, as were his parents, at least one sibling, and his first wife; and the parish register of the chapelry includes the names of most of the individuals mentioned in the Felling section of the autobiography, for instance, William Yellowly, the schoolmaster <**5**> [Heworth burial register, 19.1.1795], Sopwith Morley, the engineer <**11**> [Heworth baptism register, 26.7.1795], and William Anderson the banksman <**28**> [Heworth baptism register, 10.4.1801]. In 1834 Heworth was described as having "one corn mill, three public houses, and a few shops", also a parish school house built in 1815 - too late for A.E.'s schooling [Mackenzie 1834, 1:17].

On the occasion of drinking with the "wise man", A.E. had just had an adventure with a would-be suicide on Hebburn Fell, on the road back from South Shields <**53**>. A.E.'s mother had been at service, before her marriage, "at Powders Close, near Hebron quey" <**2**>. As late as 1834, Hebburn Quay was described as merely "two public houses, a farmstead, a few cottages, and a ferry-boat" [Mackenzie 1834, 1:16]. "Powders Close", which no doubt A.E. pronounced as 'Pooders Close', was Poulters Close, a locality on the river half a mile below Bill Quay, near the later Pelaw Main staithe, and near the "well-known creek and cottage called Lizzie Moodie's" [Mackenzie 1834, 1:26]. As shown on the 1856 OS map, it comprised a small group of buildings near staithes: the site is now grassed over and the name forgotten. Finally, A.E.'s childhood boating adventure went wrong a little further down-river, at Jarrow Quay <**16**>.

Beyond Tyneside

A.E.'s travels were in a circumscribed area. His parents had travelled a little more widely. His mother originated from Matfen Moorhouses, in Stamfordham parish, in Northumberland, 14 miles almost due west of Newcastle, and moved to Hebburn, before marrying at Felling. His father originated from Langlee in Netherwitton parish, also in Northumberland, 18 miles NNW of Newcastle, but moved to Kibblesworth, only "a straggling village" eighty years later [Mackenzie 1834, 1:154], in Lamesley parish, in County Durham, about 4 miles due south of Gateshead (as it then was), before settling at Felling <**2**>.

Although A.E. did not travel far, he was not unaware of the wider world. He read newspapers, even London ones <**35**>. Nearer home, he mentions various places in the North East: Allendale in the SW of Northumberland <**31**>, Morpeth and Alnwick north of Newcastle <**29,48,63,72**>, Tweedmouth and Spittal further north, at that time in North Durham but after 1836 in north Northumberland <**66**>, Newburn a few miles up-river from Newcastle <**10**>, and Ravensworth (Castle) in County Durham, south of Gateshead <**47**>. A travelling salesman operated in Northumberland and "the borders of Scotland" <51>. A.E. mentions "Braugh Hill fair" in Yorkshire <24>, "Hull river" <45>, Yorkshire again <48>, and a Manchester waggon <27>. Outside England, he mentions Edinburgh <52> and Ireland <9,23,66>. Finally, A.E.'s brother served in Barbados <23>, and A.E. chats to a pieman who has allegedly returned from Van Diemen's Land, that is, from transportation to Australia <42>.

COMMENTARIES

THE TECHNOLOGY AND MANAGEMENT OF CONTEMPORARY TYNESIDE COALMINING

Staithes and landsale

In A.E.'s day, most of the coal mined in the North East, hence most of the coal mined at all the Tyneside collieries at which A.E. worked, was shipped to London and other ports of southern England. In earlier times the coal had been tipped, at the river bank, into 'keels', that is, small boats, which carried it down-river to the ships waiting in deeper waters; and this had been particularly the case when the more important loading-points or staithes had been mostly above Tyne Bridge. But during A.E.'s working life this all changed. Many new mines were opened and old ones revived, and either these were situated down-river from Newcastle, or else their waggonways were so arranged or re-arranged that their coal reached down-river staithes. Moreover, new arrangements at the down-river staithes allowed the coal to be tipped directly into the ships, thus doing away with the keels. In A.E.'s lifetime the keels declined in numbers, from several hundreds to a handful, to the distress of the keelmen. A.E. refers to two staithes, the Sheriff Hill Colliery staithe at Redheugh, above Gateshead, which he visited as a child <**12**>, and the Coxlodge Colliery staithe or 'spout' near Wallsend, down-river from Tyne Bridge, where c.1820 he embarked, for a meal and a brief voyage down-river, on a coal ship, the 'Alexander' <**81**>. Apart from a reference to a single shipment of coal from Felling in extraordinary circumstances <**49**>, this is his only mention of a ship, and he nowhere mentions keels or keelmen. He worked at only one colliery loading coals above Tyne Bridge, Benwell, and this only for a short period, hence perhaps his lack of interest in keels.

A limited amount of the coal mined in the North East was consumed locally, at the mine or elsewhere. Some coal for outside consumers came from 'landsale' collieries not shipping their coal. In the late 1790s, A.E. worked for a short period at a colliery at Gosforth which was then landsale, repairing a railway which took the coal to a main road, from which point it was presumably carried away in carts <**29**>. In the 1800s a Felling pit which had previously worked as seasale changed to landsale, probably because it was almost worked out <**49**>.

Surface waggonways, waggons, and waggonway wrights

From the mine to the staithe, the coal was transported in waggons travelling on waggonways. Felling, Benwell, Percy Main and Walker collieries were near the river, and the waggonways from their various pits were of moderate length, probably the longest being a one-mile waggonway from the furthest inland Felling pit. But the waggonways from the other collieries A.E. worked at were much longer (although not as long as some earlier waggonways). Whereas in an earlier period coal mined in the Kenton area had been shipped above-bridge, by the later 1810s, when A.E. worked at Fawdon and Gosforth/Coxlodge, the coal was moved on waggonways which passed north of Newcastle and then ran down to the river near Wallsend, the longest distance being nearly five miles [Lee 1951, pp.165-172]. But not all this system of waggonways was new, since it linked up with an earlier system nearer the river. However, it appears that the new waggonways were completed before A.E. arrived in the district in 1814. Four years later, he moved to the recently opened Backworth Colliery, whose waggonway to a staithe near Percy Main was also nearly five miles in length, but also appears to have made some use of earlier way [Craster 1907, pp.27-8; Lee 1951, pp.190-1; Lewis 1975, p.114]. A.E. began work within a few days of the first Backworth coal being shipped, therefore he did not participate in the initial construction of the waggonway.

At Felling, A.E.'s father "rode the first waggon" at the opening of three pits between 1779 and 1799, most probably in the 1780s <2>, and therefore must have had some responsibility for the construction of these waggonways, being perhaps chief waggonway wright. As a child, A.E. worked with his father on a waggonway at Felling Shore, presumably one of the three mentioned above <20-22>. A.E. describes the opening of a waggonway from a fourth Felling pit in 1810 <50>, and as a fully-fledged waggonwright himself by this date, he may have had a hand in building this way before he left Felling in 1809 and after he returned in 1810. He certainly built in 1810 a complete waggonway, 345 yards in length, leading to the river (but only by joining an existing line) from a Felling pit which had reverted to landsale but whose coal, for some reason, had now to be shipped <49>. This waggonway followed the line of a previous one which had been taken up, but was nevertheless built with great speed, for a wager - A.E., working with twelve assistants, claims to have built it in two and a half days. Apart from these two instances, A.E. does not specify his working on surface waggonways, although no doubt he did this regularly.

Ideally, and therefore normally, waggonways were constructed in such a way as to lead gently down to the river bank. The waggons loaded with coal - travelling in an earlier period singly but from the 1790s sometimes in twos (hence, perhaps the running away of two waggons which A.E. witnessed as a child <**12**>) - ran down the waggonways, their descent controlled by a driver operating a handbrake or 'convoy'. In the later period brakes were doubled and their design improved [Lewis 1975, pp.199-201]. A.E. claimed that his father was "the first that put two convoys upon the pit waggons" <**2**>, but while this statement probably does refer to surface and not underground transport, no date is given, so the claim cannot be critically considered. A horse walked behind the descending waggon(s) and then was used to pull the empty waggon(s) uphill, on a second parallel track, the 'bye way'. One danger on waggonways was that of waggons 'running amain', that is, running downhill out of control, particularly if they were loaded: A.E. saw this happen on Sheriff Hill Colliery waggonway, and correctly (crede experto) blocked the track to throw the waggons off and across the bye way into a ditch <**12**>.

Where the gradient was steep, from the 1780s inclined planes (or 'balanced inclines') were introduced [Lewis 1975, p.147], and A.E.'s modest claim that his father introduced them underground at a particular Felling pit, without date given but before he retired in 1799, is worth consideration as the earliest record of their use, surface or underground, at any North East colliery. It has previously been supposed that their earliest introduction was on the surface at Benwell Colliery, by the viewer, Thomas Barnes, in 1798 [Galloway 1898, p.318]. But since Benwell and Felling were owned by the same family and since Barnes was head viewer at both, their earlier introduction at Felling is very plausible. The inclined plane worked by connecting the descending and ascending waggons to the opposite ends of a rope that passed around a pulley at the top of the incline, with the result that the descending full waggons pulled up the ascending empties. In the 1810s A.E. laid three "inclines" at Percy Main <**67**>, but these were probably underground.

Waggonways were not entirely confined to collieries or carrying coal. The Sheriff Hill waggonway had some "light waggons" on it when A.E. acted to save them, and these were perhaps carrying colliery timber <**12**>. A.E. in 1809 built, or helped to build, a waggonway at Southwick, where there was no coal, and this was perhaps to convey estate products and requirements part of the way to and from nearby Sunderland <**46**>. A.E.'s father, perhaps in 1799 (there is some doubt about this date), built a "wood railroad" at a Gateshead ropery, and this, significantly, had a waggon which moved along the rails by means of "men turning handles upon the waggon, working tooth and pinion wheals" <**18**>. This appears to have been an early version of the gangers' rolley (frequently seen on comedy films), and it represented another attempt to move waggons on rails by mechanical means, admittedly in this case for a very limited distance. In 1821, the Backworth surface waggonway began to haul waggons on an endless rope, using stationery steam engines - powered endless rope was to become a form of haulage later universally adopted underground and still in general use today. But A.E. had most probably left Backworth before this system came into operation.

Between 1799 and 1821 momentous developments affecting waggonway haulage occurred. In 1813, a Blenkinsop steam locomotive was tried out on part of the Coxlodge waggonway, with a measure of success: this locomotive had pinion wheels which ran on a rack-rail, so the experiment required re-laying the track [Galloway 1898, pp.377-383]. (The first Blenkinsop locomotive had been employed at a colliery at Leeds owned by the Brandlings, the owners also of Felling and Coxlodge collieries and therefore A.E.'s past and future employers.) It appears that Blenkinsop locomotives were used along the upper half of the Coxlodge waggonway until 1814, and then again after 1817 [Lee 1951, pp.170-172; Lewis 1975, p.115]. A.E. was working at Gosforth/Coxlodge in 1817-1818, must therefore have been acquainted with the Blenkinsop locomotives, and perhaps laid some rack-rail - but he mentions neither the locomotive nor the rail. Meanwhile at neighbouring collieries other inventors were experimenting with designs for steam locomotives. A locomotive moving on smooth rails was designed by William Hedley, built at Gateshead, and operated at Wylam Colliery in 1812-1813; while at Killingworth Colliery George Stephenson had his first locomotive operating, also on smooth rails, in 1814 [Flinn 1984, pp.153-9]. Stephenson locomotives, to some extent redesigned, soon spread, for instance to Coxlodge waggonway in 1818, but they had apparently not reached Backworth before A.E. left there, and it is not known whether they were used at Walker during his stay at that colliery. A.E. began his autobiography in 1823, by which date he cannot have failed to have observed Stephenson locomotives in action on colliery lines, and the autobiography was apparently still being worked on in the later 1820s, that is, in a period when the building of public railways was being publicly discussed. Yet A.E. makes no mention, at any point in the autobiography, of the development and use of steam locomotion, and this despite his at least approximate contact - and it may instead have been a close contact - with the Blenkinsop locomotives and track at Coxlodge.

The work of a waggon and waggonway wright included the building, repairing and maintaining of waggonways (it is curious that the relevant volume of the recent history of the British coal industry almost completely ignores the waggonway wright [Flinn 1984]). A.E. speaks of his father learning to work "with the Axe and Saw" <2> and of himself, when an apprentice, working "with the Axe and adze and augers" and also "finishing the way and working with the adze, right hand and left hand" **<17,18>**. The basic skills of the eighteenth-century waggonway wright, as also in his parallel capacity as waggon wright, related to wood-working. The sleepers and for long the rails were of wood, as were the supports for inclined planes and other devices on the way; and when A.E. is found making 'frames'for a new pit <74>, probably the pithead supports for the winding mechanism (and not the track [Lewis 1975, p.136]), these too would be of wood. But wood decays, and also splits with heavy usage: the sleepers on a waggonway might last ten years but the rails had originally to be renewed every year or two, and even after the improvement of the 'double rail' was introduced, whereby only the top rail had to be quickly renewed, the bottom rail lasted only from four to seven years [Lewis 1975, p.166]. Thus waggonway wrights were kept busy maintaining existing way, as well as building, when necessary, new way.

A.E.'s first contact with the way, when a child, was by cleaning it (actually underground) <**17**>, that is, clearing away the coal debris that inevitably fell from the waggons and built up on the inside of the rails and between the tracks; and he no doubt supervised similar cleaning when an adult. Maintaining a waggonway called for regular inspection, perhaps especially at switches and gates, and regular minor repairs. A.E. does not describe his maintenance work in any detail, but we may assume that it included sawing and shaping timber, leading and replacing sleepers and rails, leading and replacing ballast, and perhaps even cleaning ditches and repairing fences [Lewis 1975, pp.218-220]. When A.E. went to a hardware shop, he bought a gimlet, a "handsaw file", and a pair of pliars <**51**>: no doubt he kept these, with other tools, in his "tool chest", which was apparently too heavy to be conveyed by hand <**74**>.

It is likely that there were other waggonway wrights employed at each of A.E.'s collieries. In 1792 a dozen River Wear collieries employed 100 waggon and waggonway wrights, and in 1809 Washington Colliery employed six wrights and five labourers [Lewis 1975, p.218], although in 1812 Temple Main Colliery claimed to employ only two underground waggonway wrights (and Elswick Colliery none, which cannot be correct) [Flinn 1984, p.332]. The Heworth parish register records the names of other Felling waggonway wrights who were contemporaries of A.E. and his father [Heworth burial register, 1798,1799,1802]. A.E. does not state that he shared jobs with other waggonway wrights, but he certainly sometimes had had labourers to assist him. At Percy Main in 1814, he lost his contract for the way when he was "cut out by One of my Men" <**67**>.

On Tyneside from about the 1760s, wooden rails were made to last longer by having iron plates nailed on them, at first at switches and curves, later rather more extensively; and switches and points were always of iron, at least partly [Lewis 1975, pp.168-9,179]. A.E. claimed that his father "invented the Duble switch" <**2**>, but it is not clear what this refers to. The invention of the iron rail must have revolutionised the work of waggonway wrights. But the iron rail did not reach the North East until the 1790s (when it was first laid on stone sleepers), and did not sweep away wooden rails until the weight of steam locomotives made its use essential. In 1807, a contemporary reported that Walker Colliery was "gradually giving up their wood Roads and introducing Iron, and they make them piece by piece, keeping up the old Waggons, and running them partly on wood and partly Iron" [Lewis 1975, p.295]. This fits nicely with A.E.'s casual notice that the John Pit at Felling, opened in 1810, had "951 yds of Cast Iron railway" <**50**>, this being presumably the earliest Felling waggonway of this material. (In 1809 the Felling owners considered a report which compared the cost of laying "one Yard of Metal way" with that of one yard of single wood way [NCRO, Watson papers, 3/31/25].) By the 1820s, A.E. must have been building waggonways only with iron rails, and by this date mainly wrought iron rails - but he offers no comment on these changes. (It is a moot point whether his once-only description of himself as a "Mackenick", when writing to the newspapers in probably 1801 <**35**>, signified any awareness on his part that the work of the waggonway wright was changing from largely working with wood to increasingly working with iron.)

Waggonway wrights were also waggon wrights, that is, they built and repaired waggons. A.E.'s father "hired himself to make waggons and waggon way", he "wrought at waggons and waggonway all day", and was "the first that made the waggons for Old Faud Colliery" <**2**>. Since waggons were made on the surface, they provided daytime work, in contrast to the underground tasks which often had to be done at night. When an apprentice, A.E. used his expert knowledge to detect and mend a broken axle on a road waggon <**27**>. He does not mention waggon-making in relation to his early decades of employment, but at Fawdon in 1814 he had to make "12 Chaldren Waggons" and then "was repairing", perhaps waggon repairing <**68,74**>; and when he was taken on at Gosforth it may have been specifically to renovate some waggons which had been bought from another colliery <**74**>. (A.E. at first describes his trade at Fawdon and Backworth as that of "waggon wright", but it is clear from the later text that he was only using this term as a shorter form of 'waggon and waggonway wright' and that he was not limited to making waggons <67,68,77>.) The design of waggons did not change greatly during the period of the autobiography, and they continued to be constructed mainly of wood; but iron wheels had replaced wooden ones by the 1780s [Lewis 1975, pp.197-200]. Around 1820, we find A.E. going to Hawkes' iron foundry in Gateshead, to "order some waggon wheals" <**45**>.

Ever since Stephenson built his first locomotive for Killingworth Colliery, the majority of locomotive-using railways throughout the world have been built to a 4'8"-4'8$\frac{1}{2}$" gauge. Before the 1820s, Tyneside collieries used a variety of gauges, and the gauges at most of A.E.'s collieries are not in fact known. However the Coxlodge waggonway used a gauge of 4'8", in order to link with a waggonway system which had been operating that gauge for many decades, perhaps from the 1760s; and in his turn Stephenson had to follow this gauge since Killingworth linked with the system [Lewis 1975, pp.181-3]. Thus, working at Coxlodge, if not earlier elsewhere, A.E. built track of a gauge later to be worldwide. Finally, it is to be noted that, writing in the 1820s, A.E. regularly describes waggonways as "railways", and once uses the term "railroad", although not for a colliery waggonway <**18**>: these terms were unknown in the North East until just before 1800 and they only came into general use with the spread of steam locomotion [Lewis 1975, p.135].

The pit-head

The coal was loaded on to the waggons at the surface, pit-head, or 'bank' of the mine, in the various ways illustrated in well-known drawings of the 1830s [Hair 1844]. But A.E. does not refer to pit-head transactions other than those connected with the drawing of the coal up the shaft and immediately related activities. (Apart from mentioning a nightwatchman at one Felling pit <**30**>.) During the original sinking of a mine, and when no steam engine was available, descent and ascent within the shaft were effected by various forms of capstan, winch or 'gin'. Power was supplied by horses, and the gin operated a rope that wound or unwound on a drum, before passing over pulleys and down the shaft [Flinn 1984, pp.99-100]. When a child, A.E. saw a pair of 'crabs' or horse-gins drawing machinery up the shaft <**11**>; and later a crab was used to rescue workers left hanging in the shaft when the steam-engine broke down <**28**>. In 1812, at Felling, the working shaft used a steam-engine for drawing coal, but a horse-gin for letting down and drawing up workmen, and also on those days at the weekend when no coal was drawn; while the other shaft had only a gin [Hodgson 1813, p.6].

By A.E.'s day, at the Tyneside sea-sale collieries regular movements up and down the shaft were normally powered by steam-engines, although these sometimes doubled as pumping engines [Flinn 1984, pp.101-3,121-8], as was apparently the case at one Felling pit <**28**>. (Felling colliery had employed a steam-engine from its start, in 1776 [Galloway 1898, 1:294]). The engine normally stood some distance away from the shaft, and if, as A.E. states, part of the Felling engine crank fell down the shaft, it must have flown through the air when it broke off <**28**>. A.E. supplies details about the winding capacity of this engine. An engineman or 'brakeman' was said to "brake an Engine", and one engineman was only "a Lad" <**28,66,70**>. The anecdote which A.E. relates about the "One pumping ingine" at a Felling pit, apparently the Discovery, mentions various technical details, including reference to a 'scoggan', and this almost certainly identifies the machine as a Newcomen engine, a type of engine which at this date - the 1800s - was still widely in use in the North East <**39**>. Here the engineman is called a "Plugman", and he is assisted by a "fire man" who wheels coals in, while in overall charge, presumably of all the steam-engines at Felling, is a "working wright". But supervising the repair of pumps was an "Ingineear" <**11**>. One Felling pit had "two main ingines" pumping by the early 1790s and A.E. supplies the dimensions <**11**>; but at the Discovery pit there appears to have been only one pumping engine, and this not fully employed, at least until the pit "holed the waste" in 1807 and was partly flooded <**39**>. In 1818 another "engine" was installed at an unnamed Felling pit at High Heworth <**76**>. A.E. shows knowledge of the technical details of the various steam-engines working at Felling in his time there. He also refers to the underground parts of the pumps and describes in some detail their operation, in particular how the water was collected and stored for pumping <**28,43**>. He later refers to an "Ingine Wright" at Fawdon and to the Heaton Colliery pumping engine <**70**>, to "Brakemen" at Gosforth <**76**>, to the momentary fear that, when water broke into a Backworth pit, "the warter was beating the engine" <**82**>, and to the pumping situation at Walker Colliery, where the viewer "got 2 Ingines and 2 pits Opened to the botum" <**83**>. Returning to the winding technique, one of the two shaft accidents narrated by A.E. and involving himself came about when the rope "serged on the rowel", that is, mis-wound itself above <**33**>. The "New frames" which A.E. made - or perhaps helped to make - at Gosforth in 1817 were probably the pit-head winding supports, which were still of wood <**74**>.

The workman who stood at the mouth of the shaft and controlled the winding operations, by signalling to the engine-man, was the 'banksman' (a term still used). A.E. several times refers to a banksman <**2,28,32,33**>. A.E. states that on one occasion he called up the shaft from the bottom, and the banksman, in bed with his wife, heard him <**32**>. But it takes an effort to believe either that A.E's voice was quite so stentorian or that the banksman's house was so near the shaft.

Pits, seams and shafts

By A.E.'s day, Tyneside collieries normally worked, at any one time, more than one pit, although all the pits would be in the same locality (the coal-working of an area having been been leased from a landowner). However, because of the system of underground working, pits had a relatively short life, often only twenty years and sometimes ten years or less. Underground working techniques were much modified during A.E.'s lifetime, partly in order to extend the life of pits.

A.E. names six pits of Felling Colliery, but the details need to be disentangled from the narrative. A.E.'s father was at the opening of the Ann, Venture and Discovery pits, at unstated dates between 1777 and 1799, and at the opening of the John pit in 1810 <**2,50**>. The pit he was working in - presumably an early one since it was " Wheare the Felling Colliery was woon" - when A.E. had his first colliery adventure about 1790 is unfortunately unnamed <**11**>, but he escaped death in the Ann pit shaft in 1799 <**2**>. Around 1789, the schoolboy A.E. played games around a shaft of Hollihill pit <**6**>. In the early 1790s, A.E. began to work with his father in the Venture pit, where William Rogers was furnaceman, and a later episode naming the same man presumably related to the same pit <**14,19,24**>. When apparently he was already a young man and therefore not before the later 1790s, A.E. had a "singlar escape" in the Ann pit shaft, an escape which involved William Anderson the banksman <**28**>. Four later anecdotes relate to A.E. working in the Hollihill pit (or "Hillpit") <**31-4**>, and two involve the banksman there, again William Anderson, who must therefore have transferred from the Ann pit. One episode occurred shortly after A.E. married in 1798, but another anecdote names the viewer as "Mr Straker", whereas the earlier Felling viewer was Thomas Barnes <**11,62**>, who died in 1801. Therefore it seems that A.E. worked at the Hollihill pit from at least 1798 and through at least part of the 1800s, although of course his work may not have been limited to this pit. By 1810, the Hollihill pit had been reduced to landsale <**49**>, which probably means that it was closed shortly thereafter. Felling anecdotes later in the autobiography name Straker and therefore apply to 1801-1811 <**39,43,49,50,55**>. In 1807, the Discovery pit was partly flooded <**43**>, and a pumping engine episode "before the Felling pit holed the waste" presumably applies to the Discovery pit before 1807, while another reference to this pit seems to date before A.E.'s marriage in 1798 <**30,39**>. Finally, it is stated that the furnace drawing air through the Hollihill pit, apparently in the 1800s, "was at the Roddney pit" <**32**>. It can be concluded that the Hollihill pit was working from at least about 1789 and continued working during the earlier 1800s; and that the Venture pit was working in the early 1790s, the Ann pit in the later 1790s, the Rodney pit at some date during the 1800s, and the Discovery pit from before 1798 up to the later 1800s. However, the fact that A.E.'s father was at the opening of the Ann, Venture and Discovery pits indicates that they may have been opened long before the dates given above, and the Venture and the Discovery, if their names meant anything, may well date back to the opening of the colliery in the late 1770s. It looks as if Felling in the 1790s and 1800s had at least four pits working at the same time, although their working lives over the four decades may have been to some extent staggered, particularly if they were opened at different times, perhaps in pairs.

Since all these pits worked the High Main seam, which was exhausted at Felling by 1811 [Galloway 1898, pp.391-2], some may have been closed during the 1800s and others been partly closed down. The

Hollihill ceased to be seasale before 1810 <**2,6,11,14,19,24,28,30-2, 34**>. In 1801-3 the Venture pit was sunk to the Low Main seam [NEIME 1885, p.23]: A.E. seems to say at one point that the Venture worked the Low Main in a period before 1799 <2>, but this must be a slip. Whether Low Main coal was ever worked to any extent in the Venture pit is uncertain, since a contemporary source implies that Low Main coal working first took place in the John pit [Hodgson 1813, p.5]. In 1806 the pillars were being robbed in the Venture pit [NCRO, Watson Papers 3/31/1], which probably indicates that High Main coal-working was coming to an end there; yet in 1809 the Felling Colliery stock included ropes at the Venture pit, hence the shaft must have been maintained [NCRO, Watson papers, 3/31, p.225]. In 1809 the Ann pit's underground stables were still in use, all the horses there being stock-listed by name [NCRO, Watson papers, 3/31, p.229]. The John pit, working the new, deeper Low Main seam, was opened in 1810 <**50**>, and the William pit (not mentioned by A.E.) a litle later [Hodgson 1813, pp.5-6]. These two pits were apparently the only Felling pits working in 1812 when the disaster occurred, twelve months after A.E. finally left the colliery - he does not mention working in either of these pits and perhaps never did.

The newer colliery at Percy Main, which A.E. went to in 1814, worked two pits <**56,58**>, the Percy, opened 1802, and the Howdon, opened 1807 [Galloway 1898, 1:390]. A.E. does not name or number the pits at the other collieries he or his father worked at (Old Fold <2>, Benwell, Fawdon, Gosforth/Kenton/Coxlodge, Backworth, Walker), or at the other collieries he mentions (Sheriff Hill <12>, Jarrow <**62**>, St Anthony's <**62**>, Heaton <**70-1**>).

Again by A.E.'s day, a Tyneside sea-sale colliery normally used two shafts at each pit, in order to provide improved ventilation. One shaft, the down-cast shaft, drew the air in and carried it down, and the other, the up-cast shaft, passed it up and out, after it had made its way through the workings between the two shafts. An alternative arrangement, whereby two coal-working pits linked up and shared the ventilation, may have been the case at Percy Main, since A.E. speaks of connecting doors between the colliery's two pits <**57**>. A further complication is that even when only one of the two shafts of a single pit was coal-drawing, the other was sometimes known by a separate name, perhaps because it had at one time been coal-drawing - for instance, the Hollihill and Rodney pits at Felling <32>, and later the John and William pits, the latter being the up-cast shaft [Hodgson 1813, p.6]. To create a brisker flow of air, it was normal to have a brazier or furnace heating the up-cast air, the furnace being placed either at the bottom of the shaft (in which case the air was often fed into the shaft higher up, so that gases did not pass over the furnace) or at the top (in which case the air was conducted into a chimney built over the furnace, to increase the upward draught). Coal was normally drawn up the down-cast shaft.

About 1789, the Hollihill pit at Felling had a 'cube and fire', that is, a brazier and chimney at the top of one shaft <**6**>. Later, this pit had a furnace at a second shaft which was known as the Rodney pit, apparently an underground furnace, since an explosion was feared if gas came into contact with it - hence A.E. saw to it being extinguished <32>. The furnace at the Venture pit at Felling was certainly underground <**14,19**>, and when two non-miners saw it, "the large furness Reminded them of Hell" <24>. At Percy Main Colliery, there was a furnace at the bottom of the Howdon pit. When gas invaded this pit, A.E. adopted a different strategy, deliberately exploding the gas as it passed over the furnace, apparently so that the increased circulation of air would disperse the remaining gas circulating in the mine and avoid an explosion underground which might spread to a reservoir of gas <57>. (The gas "blasted 6 times" in the shaft, a very dangerous happening, and the strategy strikes me as dubious, although A.E. says it was later much commended by the eminent viewer, Buddle.) The furnace was kept going day and night, hence A.E., although on night shifts during which the pit was almost deserted, nevertheless still met the furnaceman <**14,19,57**>.

Coal came up the working shaft in 'corfs' or large, round, wickerwork baskets (made by corvers <**18**>). On one occasion, very badly burned men came up the shaft bundled into corfs <34>. (If dead, they would probably have come up, like the pit ponies, in a horse net <**24**>.) The corfs were hung on a hook at the end of the winding rope, and this hook could be inserted into a chain immediately above it, to form a loop. Miners 'rode' in the shaft, up and down, on the rope, and/or in the hook or the loop. That is, they straddled the hook or loop with their legs, or stood on the hook and held the rope above, or simply clung to the rope with their hands and twisted their legs around it. A.E.'s father came up on a hook which broke the moment he was pulled to safety at the side of the shaft by the banksman <2>. On this occasion, two men rode in the hook, and A.E., then a boy, held the rope above; on a later occasion, A.E., now a youth, rode in the loop with another individual, and eight boys clung to the rope above; on a third occasion, two men rode in the loop, and A.E., although an adult, held the rope above **<2,28,33>**. The danger is obvious. A.E. notes, quite casually, that on another occasion "One boy fell down the pit in Coming up" **<61>**. When the rope surged and A.E. was left "hanging in the shaft" for over one hour, apparently several hundred feet from the bottom, bundles of hay and jackets were laid at the foot of the shaft, surely a gesture rather than a hopeful attempt to break a fall from that height **<33>**. It was stated in 1814 that a horse-gin was used to let down and pull up workmen at Felling (perhaps this gave a steadier if slower ride) [Hodgson 1813, p.6]. But on the only occasion A.E. mentions the man-winding power source, it was a steam-engine that was in use **<28>**, and it is likely that man-winding by horses was increasingly eliminated during A.E.'s working career. Perhaps because the risks were so obvious, falls in the shaft by those travelling up and down were in fact not a very common happening, and did not contribute much to coalmining mortality. As with some other aspects of mining, experience overcame what seemed to the 'surface-lubber' almost suicidal and wholly terrifying practices.

Underground waggonways

At the bottom of the shaft was a low scaffold <28>, which received descending corfs and men. Coal was brought to the shaft in waggons drawn by horses. Horse-height passageways led away from the shaft in various directions, and one led to the stables. At the Venture pit at Felling these were located 60 yards from the pit bottom and contained seventeen or more more stalls for the 'galloways', that is, ponies or small horses **<24>**. Horses were valuable [Flinn 1984, pp.96-7] - A.E. was once congratulated on saving "all the Men and Horses and the Colliery" <57>. Stock-taking at Felling included listing the horses individually, each by its name and description

[NCRO, Watson Papers, 3/31, p.229]. When a Backworth pit was threatened with flooding, preparations were made "for Drawing away the horses" <**82**>. A.E. mentions several stablemen or "horse keepers", one of whom goes to Yorkshire to buy horses for the colliery <**28,33,34**>. Another stableman died of burns, apparently received 'in-bye', that is, in the coal-working area, where it may be doubted whether he normally went, but which he seems to have visited because he was sceptical about A.E.'s report of an encounter with a ghostly figure - the 'ghost' turning out to be of himself <**34**>.

As a waggonway wright, A.E. regularly goes underground to repair "waggons and way" <**14,19,34,56**>, the term 'way' indicating the rails laid down in the fairly high passages or tunnels which approached the in-bye coal-working areas, rails along which horses pulled waggons. ('Way' also means 'the passage of waggons along the rails', hence "the way Started" <**62**>.) Since a waggonway wright generally did his work during the maintenance, non-coalworking shifts, at night or at the weekend, he did not normally encounter waggons and horses on the way, but A.E. mentions meeting a maintenance man's timber waggon and horse <**34**>. A.E.'s working notes detail the laying of long lengths of way in the Hollihill pit at Felling <32>. As the coal-workings advanced, so did the approach tunnels, requiring the laying of additional way, which had to be torn up again and the rails removed when the working area was exhausted; and new tunnels to new areas regularly required new way. In addition to this work, the underground waggonway, like the surface one, required continual maintenance and repair - A.E. specifies his father and himself going underground because "at the 5th Siding the switch wanted repairing" <14>. A reference to "repairing the horse road Called penning", after which A.E. "retreated to the waggonway" <**32**>, is slightly mysterious but probably refers to an advanced passageway without rails along which a horse simply dragged a corf, perhaps on a sledge, and perhaps on a wooden floor. Waggonway wrighting was no doubt hard physical labour, but even underground it was seldom as exacting as on the occasion when A.E. and his son "had to work on the way half leg depth of Warter for 12 hours" <**56**>.

As part of the waggonway, the waggonway wright built and maintained inclined planes underground. As already noted, A.E.'s claim that his father was the "first that made the Self acting incline of waggonway in the Ann pit of Felling" <**2**> is possibly of some historical interest (it certainly challenges the statement that the first inclined plane underground was installed in 1800 <Galloway 1898, 1:319>). A.E. himself "Laid 3 Inclines" at Percy Main, probably underground, and he reports that later, at Backworth, "We had Lentheaned the plane by 2 pillars and ... set Double doores above the Incline wheal" <**67,78**>.

The coal was brought to the waggonway from the coal-working faces in corfs, and these were lifted on to the waggons by means of small hand-cranes. The waggonway wright was responsible for maintaining these cranes and moving them on as the workings advanced. At Felling in 1799 A.E.'s father was "Imployed in the Sunday night making some new road and seting the Crain in the Ann pit" <**2**>, and at Percy Main in 1813 A.E. and a partner were "shifting the Crain 2 pillers" <**62**>. Nine-tenths of the time a waggonway wright spent underground must have been occupied with the waggonway duties referred to above, but he could also assist with other underground tasks, especially in emergencies; and since A.E.'s anecdotes about underground work concentrate on emergencies, we hear of him assisting with occasional or one-off jobs - putting up doors and brattices after an explosion <**14,78**>, plugging a flood-hole <**82**>, helping with a percussion experiment <**61**>, and swimming into a flooded section to check the extent of inundation <**43**>.

Underground coal-working

If a coal seam be first imagined as a meat slice in a sandwich, then the problem of coal-working is that of removing the meat given that (a) the upper bread slice is often more than one hundred times thicker than the meat, and (b) the scale is such that the men work almost solely within the meat slice. To prevent the workers being crushed within the coal seam, the roof must be supported while the coal is being removed from the working area (a temporary measure, since eventually the roof will be forced down - no bad thing, because this relieves the downward pressure elsewhere). In A.E.'s day the supporting was done mainly by leaving 'pillars' of coal, although in the later decades there was some experimenting with built-up barrier-supports of stone. The pillars left as the work advanced were very large, but when the coal-faces had gone forward as far as practical, the pillars were 'robbed' as the workmen retreated. Although these aspects of mining technology were of fundamental importance for the economics of the coal industry, they did not directly concern a waggonway wright and hence are not discussed in the autobiography.

The waggonway tunnels ran from the shaft towards the sections of the mine to be worked. From the waggonway, narrower and lower tunnels or 'headways' were driven into the coal seam, normally two in parallel, to provide ultimately a flow of ventilation. From these headways, galleries or 'bords' were formed to one side, or to both sides, at intervals, these being the working faces of the mine from which the 'hewers' obtained the coal as they drove the bords forward. The coal-working section therefore took on a chequer-board pattern, with standing square or oblong pillars of coal surrounded by excavated or part-excavated passageways. Most of the adult workers in a coalmine were employed as 'hewers' and worked in these headways and bords, which were the active front of a mine. Inevitably therefore many contingencies occurred in these places, some of which threatened the whole mine, so that although A.E.'s job as a waggonway wright did not necessarily take him close to the coalworking faces, his description of emergencies leads him to refer to them. Thus, in Howdon pit an emergency is created when the ventilation fails and "the foul air Came along the back headways" <**57**>. (A.E. refers specifically to 'headways' only in one other place <**61**> and instead uses the term 'bords' loosely to refer to all the narrower passageways off the waggonway, e.g. "the East board ... Droven ... 200 yds down" <32>.) We hear of maintenance men going in-bye "to Clean the boards of stones" <**78**>, that is, remove any overhead rock that had fallen or been brought down with the coal and sorted from it. Three episodes involve gas penetrating a bord, twice with an explosion, and once with a subsequent fire of "the Coal and timber", the latter presumably being the pit-props used to give temporary support to a dangerous roof <**32,34,61**>. Although in an earlier period a bord was normally so narrow that only a single man worked there, because there was not space for two to swing their picks [Compleat Collier 1708, p.42], A.E. specifies that when there was an inundation at Felling,

"George Hunter and George Riddley was in this board, 5 boards north of the Crain, Hunter on the right and Ridley on the left" <**43**>. A.E. only once refers to the actual digging out of the coal with "Pick and Mell [= hammer] and wedge" <**62**>, and does not mention the use of gunpowder to shatter the coal-face, an improvement that came in during his later career underground.

From the bord to the waggonway, the coal was conveyed in corfs, which in this period were increasingly carried on rolleys or 'trams' running on wooden boards or even rails, pushed or pulled by 'putters' who were boys or youths; and not, as previously, dragged along on sledges. Although one might have expected the waggonway wright to have had some responsibility for laying boards or tram-rails, it is not clear that he did, and certainly A.E. says little about this aspect of haulage. From the waggonway he once mysteriously hears a "tram runing up One board and down the Other" <**62**>, and when a small section of the pit is flooded among those drowned is a putter <**43**>.

The dangers of gas and water

A.E. began permanent waggonway work underground when he was aged 14. "And then again going with my farther and brother John, in a short time the fear of the pit left me. And with being in varius parts of the mine, I understood the Carrying of Air - yet never had anything to do with that part" <**17**>. Almost without exception, the Tyneside collieries, in almost all the seams worked in A.E.'s day, had to battle against gas. The "Carrying" or 'coursing' of the air was the current general solution. Between the shafts and through the whole underground workings the air was conducted along more or less a single path, in order to maintain its force and enable it to sweep out and disperse concentrations of gas. The path through the network of tunnels and passageways was maintained by closing off side-passages and cross-passages, sometimes with brick walls and other 'stoppings' or 'standings' <**32,34,61,78**>, but more often with temporary barriers (temporary because the shape of the workings was always changing), air-doors and wooden or canvas partitions known as 'brattices'. Thus the flow of the air was directed and kept moving through all the workings. For instance, "the whole Air of the pit was forced by a brick wall ... [and] the Air turned to the right hand" <**34**>.

Because the miners used naked lights in the form of 'lows', that is, candles - universally until the safety lamp was invented (an "invinsible" improvement <**80**>), and commonly even thereafter - minor explosions of methane gas or 'fire damp' often occurred. Once A.E. saw gas "as mist Coming down out of the hole where the Water had come" <**32**>, and an increase of gas in the air could sometimes be detected by the candle changing the colour of its flame - "I saw the Candle just at the firing point" <**57**>. Candles had to be immediately extinguished, and thus - "he clicked his Candle and there was no fire" <**14,32,34,57**>. But this left the miner groping in the dark. Hence the earlier experiments with forms of providing light that did not ignite gas - or were supposed not to - so that A.E. went exploring the passages of Tynemouth Castle with "2 Steelmills and 2 dried hadecks" (the latter intended to give out a phosphorescent glow), as well as with candles and a safety lamp <**80**>.

Since explosions of small quantities of methane gas were not very powerful, the gas was sometimes deliberately ignited and burnt off, and this rather casual attitude to the danger is to some extent confirmed by A.E.'s anecdote about his first experience of an underground explosion. His father carelessly lifted his candle up towards the roof of a passageway, in order to catch drops of water to moisten the clay in which the candle stood, and "the gas that had gathered above the Levell of the Air", near the roof, fired <**14**>. (In 1812, a party of visitors, including two ladies, was taken down Howdon pit to see an inclined plane and a gas blower, the latter presumably lit and flaring [NCRO, Easton Papers, 3, M.Dunn's work diary]).

However, A.E. reports instances of men burned, fatally or otherwise, when gas took fire <**14,36,61,78**>. But more serious than this immediate consequence of an explosion was the disruption to the ventilation, should the coursing of the air be short-circuited as a result of air-doors and partitions having been blown down. A.E. cogently expounded this peril to a meeting of the "Society esteblished at Percy Main for the betar ventalation of the mine", stating that if the air-doors collapse, "the whole of the Air is taken off the Intearier of the mine", and an asphyxiating gas produced by the explosion, known locally as "the After damp", then "produses death to those caught in it" <**58**>. In fact, most of the 92 deaths in the Felling disaster of 1812 were caused, not by the initial explosion, but by the action of this gas when the ventilation failed [Hodgson 1813, pp.14-15; Flinn 1984, p.136]. Even minor explosions often blew down doors and partitions, and A.E. twice records putting these up again as quickly as possible after an explosion - "We got the doors up and the Bretishes and Stopens put up with deals" <**14,78**>. On another occasion, a fall of stone had knocked down an air-door (this was in Percy Main colliery which had a particularly complex coursing system, involving 'crossings' of the air path), and before the ventilation could be thoroughly restored, gas had penetrated towards the shaft <57>. A.E. suggested a solution to the collapse of air-doors, a solution which was, he claims, taken very seriously by the viewer, John Buddle - the installation of propped-up hanging doors which by not resisting the blast would stay in place, but when the props were blown down would then fall and close, thus restoring the air circulation <**58**>.

While gas continually seeped in small quantities out of certain coal faces, it could concentrate in those areas of the mine which had been abandoned after having been worked and which, despite roof-falls and 'creeps' (the movement of the floor upwards around pillars), contained loosely packed material and air pockets. The alarm at Howdon pit when gas was found in the working passageways arose because of a reservoir of gas nearby - "you know whates is heare, 3 yds off, 18 boards and 16 piller of Creeper Waste bared off by a bord and Stoping" <57>. From such reservoirs, gas forced itself at pressure through walls, emerging as a 'blower'. In one instance a blower "hised like a Serpent" <**61**> and in another it at first roared like "the presser of Steem from a large boiler" and later alternated with water and pulsated rhythmically, leading to it being known as 'The Drummer' <**32,34**>. (In the 1810s gas from such underground reservoirs at Wallsend Colliery was led to the surface in pipes and burned off, as today happens at oil platforms [Galloway 1898, 1:488]). Gas blowers underground could be accidentally ignited, and A.E. records an early experiment in extinguishing combustion in such a case, by percussion <**61**>. (At the same colliery, at a later date, two cannon were kept in readiness underground [Galloway 1898, 1:487]).

The second major danger to the Tyneside collieries and their workers arose from water. In an ideal world for mining, coal seams would lie perfectly horizontal. But the Tyneside seams lay somewhat askew, that is, they 'dipped' in one direction and 'rose' in the opposite - in his working notes A.E. refers to one seam at one point "rising 2 inches in a yard" <**32**>. The local seams also had breaks, otherwise faults or 'troubles' - "their was a trouble in the mine of 2 feet rise to the East" <**14**>, a line of continuous strata-slip forming a 'dyke' <**14,61,62**>. Such a dyke ran through Felling colliery [Bailey 1810, p.31]. Partly because of the dip and the breaks and faults, water descended through the seams and seeped into the coal workings, sometimes in 'feeders' or streams. In general it could be dealt with by pumping, and A.E. refers to underground sumps and underground shafts or 'staples' which collected the water for pumping <**14,28,32,34**>. But when water gathered in large quantities in abandoned workings, the water, if suddenly released, could flood out a level of a mine. This might happen if workers incautiously holed into a neighbouring abandoned section or 'waste', and A.E. supplies a detailed account of such an event in one section of the Discovery pit at Felling - "holing the waste with loss of life" <**43**>. But on this occasion only a small section was flooded and only three men were trapped and drowned; and on a later occasion when a Backworth pit was threatened with flooding, "we got the hole plugged on time" and apparently no lives were lost <**82**>. It could be very much more serious if the workings holed into a wholly abandoned mine, as happened at Heaton Colliery in 1815, when 75 men were trapped and lost, an event in which A.E. took much interest <**70,71**>. To avoid this sort of catastrophe, barriers of coal of fixed width were agreed between collieries, and these were not to be worked from either side. This explains part of the alarm at St Anthony's Colliery in the 1790s when workers heard the sound of hewing - picks, hammers and wedges - coming through the coal they were working and apparently proceeding from the neighbouring colliery across the river, Felling. Convinced that "Felling would hole in a fue days" they stopped work until both collieries were re-surveyed and they could be convinced that the barrier between them was intact <**62**>. The truth of this anecdote is supported by a note in a 1806 report on Felling Colliery - "the Barrier under the River cannot be taken away, about 10 yds should be left to keep out water lodged in St Anthony's Colliery" [NCRO, Watson papers, 3/31/1]. A.E.'s working notes refer to a forward boring being done at Felling to discover how close the workings were to a neighbouring pit, the Admiral Keppel pit, which was perhaps abandoned, and also to old and uncharted pits which had once worked higher seams - "boaring feared of Old pits above the High Felling, 2 of such pits was not down to the high main" <**32**>.

The management of Tyneside collieries

The section foremen or 'deputies' are often mentioned by A.E. <**32-4,43,57-8**>. But it is doubtful if their responsibilities, which were mainly to supervise the hewers and putters at the coal face and check the ventilation, impinged much into a waggonway wright's sphere of action, although A.E. often associated with deputies in emergencies. (Of three Felling deputies named by A.E., two, Robert Stove and Martin Greenar, escaped the 1812 disaster only to be killed in an explosion in 1813; the third, Nichol Urwin, may or may not have been the man of that name who died in 1812 leaving eleven children [Hodgson 1813, pp.25,36-7; Sykes 1833, 2:76-7].)

The overman or shift foreman supervised the whole labour force in the pit at any one time, including the underground waggonwaymen - A.E. probably received from overmen, if not day-by-day instructions, at least regular directions. He says little about these regular contacts, but frequently mentions associating with overmen in emergencies, and for his good works is offered the post of overman himself (or so he says) <**33,34,43,57-8,61**>. Overmen had responsibility for the overall working and safety of the mine, so we find them going down the pit before the main shift and inspecting problem localities in-bye <**14,33**>. But when A.E. speaks at a meeting, "I was Shouted at by the Overmen in great derison, what did I know?" <**58**>. One of the overmen at Percy Main was George Cooper who was "Called the wisest man in Percy Main", and as A.E. remembered bitterly, "It was he that shouted me down" <**67**>. Presumably it was Cooper whom A.E. was aiming at when he remarked that "the Wise men of Percy Main Called me a fool - what did I know? But I judged them not" <**62**>. ("Samuel Cooper jnr., overman, Percy Main", probably a brother of George, was in 1827 a subscriber to a two-volume history of Newcastle [Mackenzie 1827, p.vi].) However, A.E. had earlier trusted a Felling overman sufficiently to consult him about a ghost, and the overman took charge of the compassionate search of a widow's house for her late husband's hidden savings <**43**>.

Cooper's father, Samuel Cooper, was "viewer" (or more likely, under-viewer) of Percy Main Colliery <**62,66,67**>. ("Sammy" is very frequently mentioned in the 1812 work diary of Matthias Dunn, who was assistant to the chief viewer, John Buddle [NCRO, Easton Papers, 3].) A.E.'s relations with Samuel Cooper were probably not improved by the chief viewer, Buddle, congratulating A.E. for proving Cooper wrong over the mysterious noises in the colliery <**62**>. Matters were further not helped by the two men courting the same woman and each claiming that the other had fathered her child <**67**>. It is hardly surprising that, although Cooper allowed A.E. to miss a shift at work in the hope that he would engage himself to the pregnant woman, he subsequently saw to it that A.E. lost his waggonway contract - "I was cut out by One of my men, by his Making Sam a preasent of a Pig 18s of valliue" <**67**>. When Howdon pit "blasted", Buddle was sent for but Cooper came with him <**57**>. Buddle or his assistant, Dunn, visited the colliery regularly, but since Buddle had very many other responsibilities, in terms of day-by-day duties Cooper was effectively viewer at Percy Main.

A 'viewer' or manager of a colliery was responsible to the owners for its efficient performance, technical and economic. He did the overall planning for the mine, sometimes in active consultation with the owners, and introduced improvements, but he also had overall responsibility for the daily routine of each pit, and to this end made inspection visits underground at regular intervals. The more successful viewers came to manage more than one colliery, and eminent viewers like Buddle were notionally in charge of many concerns; but in this case they entrusted routine duties to an under-viewer at each colliery. But in emergencies the viewer was always called out, especially to direct rescue operations; and the less eminent viewers were liable to be blamed for disasters, as Straker was for the 1812 Felling disaster (whereas no-one ever pointed the finger at Buddle).

"People seem so dreadfully prejudiced against Mr Straker so much so that he dare not appear in public"; and six months after the disaster he was "discharged from Mr Brandling's concerns" on account of "dissatisfaction at his Management of the Colliery" [NCRO, Easton Papers, 3, Dunn's work diary, pp.103,173].

When a child, A.E. was patted on the head by "Mr Thomas Barns, the Viewer" at Felling, who apparently knew Robert Errington, his waggonway wright, well enough to recognise his name; and in the 1790s A.E. assisted Barnes when he surveyed the pit <**11,62**>. Barnes, who died in 1801, was the most eminent viewer at the end of the century, and his successor in eminence, Buddle, acknowledged this by once saying to A.E. "I am quite satisfied. Mr Barns would be Correct" <**62**>.

Barnes was succeeded at Felling by "Mr Straker" - A.E. calls all viewers and owners "Mr", except when he is very cross with "Straker". A.E.'s relations with John Straker were of varying character. When A.E. saved the Hollihill pit from the blower, Straker was called out, and on arrival "Mr Straker said I was worth my weight of gold" <**32**>. A.E. is only an amused observer when Straker checks on the drawn-out activity of the pumping engine (he calls the firemen by his first name and the fireman calls him 'sir') <**39**>. But when a section of the pit floods and A.E. thinks a life can be saved by drastic action, "I went Emedtly to Mr Straker who saw the hope and Ordered 18 men to go with him directly" <**43**>. In 1810, returning to Felling from Benwell, A.E. built a "railway" from Hollihill pit to the river, to release coals originally intended for landsale but unexpectedly sold for shipping to Scotland. Straker made and won a bet that the railway would be built within two days, and rewarded A.E. and his assistants with food and drink <**49**>. In the same year, when the John pit railway was opened, and A.E.'s father was one of the company celebrating the occasion at the Unicorn Inn at Felling Shore, Straker, in a temper at another workman during the dinner playing the old man's fiddle, presumably badly, "Came down, seased the fiddle and Broke it over a Chair top" <**50**>. But when he learned that the instrument belonged to old Errington, he "Sent to Felling Hall for his own fiddle". Whether Straker actually lent A.E.'s father his own fiddle is not made clear, but twelve years later, after the old man had died, the family fiddle was repaired at Straker's expense, "the cost £1 4s". However, "Mr Strakers fine promises to Anthony proved false" when, in 1811, he gave A.E. lower wages than A.E. thought he was worth - "Strakers Coldness to Me was bace ... the Stuert said Straker was out of his mind" <**55**>. This disagreement probably occurred on 18 April 1811, the date mysteriously appended to the title of another anecdote <**50**>, and it occasioned A.E.'s leaving Felling for good (or at least during the period of his autobiography - he returned to the locality many years later). It incidentally also saved his life, because had he stayed he would almost certainly have been killed in the 1812 disaster. During the next ten years or so, A.E.'s work does not seem to have brought him into contact with Straker. But the final anecdote in the autobiography tells how Straker replaced another viewer at Walker Colliery, and how when he urgently needed a waggon wright (to make waggons for a surface railway already laid), he "Sent for me, and I was Ingaged as a more faithful servent" <**83**> - an unexpected and surprising reconciliation on the part of both parties. A.E.'s reference to Straker's repair of the family fiddle in 1822 suggests that this may have been the date of the reconciliation.

In 1816 or 1817, when his wages were unexpectedly reduced at Fawdon Colliery, A.E. was contacted by "Mr Thos King the Viewer for Mr Brandling", and in consequence moved to Gosforth Colliery <**74**>. But he soon fell out with King. He attempted to appeal over King's head to the colliery owner, and when this failed he left <**76**>. It is not certain, however, that he blamed King for the incident which produced this outcome. He mentions King once again, but only to say that at an unstated date, probably in the early 1820s, King was viewer at Walker, where he opened up two pits but then left <**83**>. Buddle's letter book contains a confidential reference to "two wild fellows that I should object to, viz., Straker and King" [NCRO, Buddle Papers 15, p.239].

In contrast to his embittered relations with Straker and King, A.E. always speaks of Buddle with great respect, and indeed his anecdotes give a very favourable impression of the most famous of the viewers. Working in Howdon pit in 1813, A.E. prevents the pit from filling with gas and blowing up, although the shaft "blasted 6 times" <**57**>.

"A man and horse went off to Walls End for Mr Buddle. None would Come down unto he came." According to A.E., Buddle was immediately told that "had it not bean for Anty Errington, wee would have bean all dead men", and he therefore summoned the whole pit workforce to the colliery office the next morning, together with "the owners", and in front of all offered A.E. the post of overman. When A.E. eloquently refused, Buddle said, "did ever man Make a more Pathitick speach towards his fellow Man?", and shook hands with him. This episode A.E. may well have considered the high point of his life, hence, if we may assume that the story had a genuine factual basis, some embroidery and inflation in the autobiography when recounting it would not be surprising. Other anecdotes, however, show Buddle in an equally friendly mood towards A.E. Buddle approaches A.E. after the latter is shouted down at a meeting, and takes seriously his suggested improvement <58>. He employs A.E. in a percussion experiment <61>. He congratulates him when he explains away the sounds that have scared other men - "Anthony, thee has Made a grand Discovery. Thee hath A Penetrating Mind Inn a Coal Mine." <62>. When A.E. leaves Percy Main, he blames Sammy Cooper, not Buddle, although it is curious that he does not refer to any attempt, similar to his later attempt at Gosforth, to go over his enemy's head, and in this instance to appeal to Buddle <67>. One year after leaving Percy Main, A.E. happens to be a bye-stander at Heaton pit after the disaster there, and when Buddle returns to the surface from an inspection below ground, A.E. is recognised and spoken to in friendly fashion by the "very miry and exasted" viewer <70>. A.E.'s assertion that he had been led to Heaton "by the Devine Speret" was probably regarded by Buddle as evidence of a side of A.E.'s character which he had perhaps been hinting at in an earlier conversation - "he said I was a perticular man" <62>. Some years later, at Backworth, after A.E. had had an absurd adventure with a donkey, Buddle visited the pit and "wated to 10 oclock to have the Storey from my own mouth ... Mr T Taylor and he got a good Larf" <81>. Finally, A.E. states at one point that Buddle said to him, "Anthony, thou shalt live by thie trade all the days of thie Life", and A.E. adds - "Which promise he fulfiled to his death" <57>. Buddle died only in 1843 and it is uncertain whether A.E. meant this remark to apply literally. If he did, then presumably A.E. was employed, after the period of his autobiography, at one of Buddle's collieries, perhaps at Percy Main again.

A.E.'s account, whatever its factuality in detail, testifies to Buddle's conscientious performance of duties that took him into troublesome, unpleasant and dangerous situations underground, and it tends to confirm that Buddle enjoyed warm relations, to an extent probably untypical for a viewer, with ordinary colliery workmen. He established a mines safety group among the workmen at Percy Main, and "all that atended the Society was treated with a super at North Shealds. To frame Love and Friendship toward Each Others welfare - this was Mr Buddles intent in doing so" <**59**>.

At Backworth, the viewer or under-viewer (apparently working under Buddle) was a Mr Oliver. In one of the two anecdotes that mention this man, A.E. volunteers to work with him in an emergency, and in the other they chat while A.E. eats, and perhaps shares with Oliver, "a Couple of beskits and a Slice of Corned Beef" <**78,81**>.

Whether it be in relation to Straker, King, Cooper, Oliver or Buddle, the autobiography makes clear that a waggonway wright like A.E. was brought into closer relations with the upper management of a colliery than was the case with the majority of underground workers. A.E. blames Straker and Cooper for direct interference with his economic prospects, and King for unsatisfactory orders, probably related to money matters. It appears that he negotiated with the viewer on long-term waggonway contracts, which is understandable if only the viewer knew the forward plan for coal-working. In emergencies, he is allowed access to the viewer and may work with him.

In fact, A.E. had contacts even higher up the colliery hierarchy. He approaches "Mr Robson the Agent" or "Stuert" of Felling about a lost soul <**43**>, and Mr. Robson later "wished me to stop in Mr Brandlings Imployment" <**55**>. "Mr. Brandling" was either C.J. Brandling, the supporter of George Stephenson in the safety-lamp controversy, or his younger brother, R.W. Brandling, founder of the Brandling Railway, both of them the chief owners of Felling Colliery and later Gosforth Colliery (a third brother, John, is also mentioned <**49**>). 'Mr. Brandling' took some interest in his colliery concerns - when in 1807 a man was rescued after being trapped underground, "he was ordered by

Mr Brandling to go no more down the pit" <**43**>. Whether Brandling provided the victim with another livelihood is not stated. Ten years later, when Gosforth Colliery required a waggonwright, "On naming that to Mr Wm Brandling, he said, where was the Erringtons. I demand one of the Erringtons, my old servents." <**74**>. But when A.E. disagreed with Brandling's viewer, and said he wished "to have a hearing before Mr Brandling", the latter refused to intervene and presumably to see A.E. <**76**>. At Percy Main, as we have noted, A.E. made an appearance before the assembled owners, to be congratulated; and this assembly is just plausible because of Buddle's influence on local coalowners, he himself, at least at a later stage, being a part-owner of a number of local collieries. When A.E. went to Backworth, he had some degree of contact with another coalowner, Thomas Taylor. Taylor joined Buddle in "a good Larf" at A.E.'s account of his adventure with a donkey <**81**>. When Taylor was organising an expedition into the underground passages of Tynemouth Castle, and looking for volunteers, "I told him I durst go first" <**80**> - but it is not clear from A.E.'s account whether Taylor himself accompanied the party underground and joined the subsequent celebrations. However, when A.E. decided to move to Walker, Taylor granted him leave to go "as he was Senier Master" <**83**>. Again, it is unlikely that an ordinary collier would have had any contact with the coalowners, and although we may accept that in the autobiography A.E. makes the most of contacts he clearly found flattering, nevertheless it is patent that A.E.'s occupation gave him a status which to some extent set him apart from the mass of colliery workers.

COMMENTARIES

A.E.'S FAMILY, HOMES, AND DOMESTIC ARRANGEMENTS

His parents

As a child and young man, A.E. lived with his parents, who eventually had four sons and five daughters. Anthony was the fourth son, born when his father was aged 46: we do not know when most of the daughters were born, but probably A.E. grew up as one of the youngest members of a large family. (See the Appendix at the end of this Commentary for the family tree.) After moving to Tyneside, probably in his mid twenties, certainly by his mid thirties, A.E.'s father lived there for over fifty years; and as far as we know, all his children remained on Tyneside throughout their lives - except George, who joined the army and went overseas, but who returned after nearly thirty years away and spent his last decade of life on Tyneside.

After A.E.'s father died in 1818, at least two and perhaps all three of his presumed unmarried daughters (Jane and two sisters unnamed in the autobiography) who had lived with him, at least in his old age, appear to have gone to live with, or near, their oldest brother, William, at North Shields. (William may have been unmarried, since in 1818 a man of his name had a child baptised who seems to have been illegitimate [CRS 1836, p.283].) The evidence for the removal of the sisters from Felling is that A.E. records spending an evening c.1820 with "my Brother and sisters", apparently at North Shields, after having had dinner aboard a ship at the request (or invitation) of "My Brother William" <**81**>. When A.E.'s youngest son, John, was baptised in 1825, the godparents were John and Margaret Errington, most probably A.E.'s second oldest brother and one of the sisters [PRO, North Shields Catholic register]. (For possible identifications of the North Shields siblings, see the Appendix at the end of this commentary.) At an earlier baptism of one of A.E.'s daughters, in 1806, the godmother, called 'Joanna' in the Latin entry, was probably A.E.'s sister, Jane [CRS 1936, p.238]. As a final instance of relatives acting as godparents, when A.E.'s youngest recorded daughter was baptised in 1831, the godmother was one 'S. Errington', probably A.E.'s much older natural daughter, Sarah [register of St Andrew's Catholic chapel, Newcastle]. A.E. mentions a brother-in-law

<35,52>, therefore one of his sisters was married, perhaps the oldest one: in the 1800s she lived in Newcastle, and may have been the Sarah Errington, married to Richard Clarke, who in 1792 had a child baptised and who was at that date living at Gateshead [CRS 1936, p.213]. A.E. had well-developed, even excessive, feelings of family loyalty. According to his own account, when he asked permission for his soldier brother George to spend a night at the family home, he assured the colonel of the regiment that George would never desert, for fear of being disowned "by father and mother, brothers and sisters" <23>.

The whole of the married life of A.E.'s parents was spent at Felling. When his father retired from his trade at the age of 67, he took a small farm <2>: it is not clear whether previous to that he had lived in one house only, or had moved from house to house. In the early 1790s, he lived in a "Cotage" <23>. A.E. praised both his father and his mother for their industry. Before she married, his mother had been in domestic service from the age of eleven - her age at marriage was 24 and her husband was aged 34 <2>. Both parents had been born in the agricultural countryside, some way outside semi-industrial Tyneside, and had met when both were employed at or near Felling. A.E. spoke warmly of his mother : she attended to him instantly when he fell into a grave <5>, she took him away from his first job when he complained <13>, and she taught him the spells necessary to make butter set <20>. His mother and his sisters churned milk for customers <20>: presumably the family kept a cow or cows, and the expenses of a large family were sustained by the earnings of both parents. The final reference to his mother, a casual one, is when A.E., as a boy in his early teens, goes to Newcastle on Fair Day to meet his mother and father <27>. Thereafter the autobiography fails to mention his mother, although in fact she did not die until 1802, when A.E. was aged 24 and was four years married. Perhaps her relations with her son had been less close after his childhood, and conceivably she had disapproved of his early marriage to an older woman.

A.E. only knew his father when the latter was in late middle-age or was elderly, but he speaks of him with warm respect and admiration. He made A.E., when a small child, a toy waggon <4>. Later, A.E. took his father's dinner to him at work <**11**>, and was sent by his father with a letter on an errand to a neighbouring colliery <**12**>. His father twice put him to a rural trade, once to blacksmithing, and once to farming <**13,15**>. But when A.E. disliked these jobs, he took him to work with him at Felling Colliery, at first part-time <**14**>, and finally, when A.E. was aged fourteen, took him as an apprentice to himself, to learn the trade of waggon and waggonway wright <**17**>. A.E. specifies what he regards as his father's achievements in his trade <**2**>, and notes that the old man was called back from retirement to share in a celebration at the colliery where he had worked for many decades <**50**>. Despite A.E.'s tributes to his parents, they rapidly disappear from his anecdotes, even those about his childhood; and once he starts work on his own his father is not mentioned for many years, not in fact until he retires and eventually, it seems, shares a home with A.E.'s family.

His first marriage

A.E.'s first wife is only briefly mentioned: she had worked as a "housekeeper" locally, perhaps for some years, given her age, and A.E. appears to have met her through the good offices of a friend of his who was courting a friend of hers <**31**>. When they married in May 1798, he was four months under 20, she two weeks over 30 [NCL, Kirkwhelpington parish register transcript]. "I got equented with her and in proses of time I married her" is his laconic statement: she was four months pregnant at marriage. (The parish register entry records that A.E.'s first child was born in October but lacks the year [DCRO, Heworth parish register], however A.E. lets slip that it was born within twelve months of marriage <**31**>, therefore it was born in October 1798.) Early in their married life he sympathises when his wife sees a spirit, to the extent of agreeing to move house from High Felling, where their first child was born, to Low Felling <**31**>. Since he "applies" for a house, he was most probably living in tied colliery housing, that is, a house was provided with the job, for a fairly notional rent. The new house was "a cottage where Lees factory is at preasent" <**44,46**>. Whereas his mother had contributed to the family income by her dairy activities, A.E. does not indicate that either his first wife or his second wife worked at other than domestic duties. Later in his marriage, A.E. once goes marketing with his wife and once goes with a neighbour <**38,45**>: on the latter occasion perhaps his wife was near her time with one of their four children - or perhaps she was sick, since she died in 1809 after only ten years of marriage, aged 40. A.E.'s tribute was cool, he and his children had lost "a Mother and honest partner in Life", and he was more obviously concerned about the loss to his young children <**46**>. On the other hand, he now changed his way of life, moving away from Felling almost instantly, at first temporarily but later permanently, and thereafter roaming from colliery to colliery in a way he had not done before, perhaps evidence of a deeper emotional upset than he could put into words in his autobiography.

A.E. was now regularly separated from his children. When sent across the Tyne to work in 1809-?1810, he "dwelled at Felling but lodged at Benwell" <**46**>. Since he probably lost his Felling house when he finally left that colliery in 1811, and since his children seem now to have joined their grandfather and housekeeping aunts, either A.E. found another house for the joint establishment, or, more likely, they all lived in the grandfather's existing house, perhaps "the small farm". His father, although now very elderly, and despite, from the age of 76, being "efected with the Rumatis at times" <**2**>, was still active. When nearly 80, we learn that he has been marketing in either Gateshead or Newcastle, although he now requires help with his market sack <**54**>. During his short stay at Benwell, A.E. lodged with Mrs Bowman, a widow, at Fenham Lodge <**48**>; but two anecdotes refer to his returning to Felling on a Saturday or Sunday <**47,48**>, which he seems to have done regularly during the next decade, no doubt to see his children. However, during the period at Benwell he also fathered a child on a woman called Jane Richardson, a daughter, Sarah, conceived nine months after the death of his wife [CRS 1936, p.251]. Although he fails to refer to this episode in the autobiography, he took the child into his household and mentions Sarah later <**79**>. The gap in the autobiography suggests shame at this breach of family and religious ideals. But the very curious fact that the recorded baptism of Sarah is almost the only bastard baptism recorded in many decades of the Catholic register must indicate that A.E. took the breach more seriously than equally errant fellow-Catholics - who perhaps made do with Anglican baptism - and may indicate the depth of his desire for reconcilation.

He finally left Felling Colliery in 1811 and moved to Percy Main, across the river, "leaveing my Aged fauther, and none to help him but 2 Sisters" <55>. Yet elsewhere he mentions meeting "a girl that had lived Servint with us" <48>, indicating that the family had had, at least at one stage, a domestic living in, perhaps a teenager, as was not uncommon in earlier centuries, even in very humble homes. A.E.'s oldest son, Robert, aged about thirteen, joined him at Percy Main to learn his father's trade, but took ill at Felling one weekend: A.E. received a letter at Percy Main and "got the Steamboat up and was at my Fauther's at 12 Clock". After recovering, Robert was bound to a painter at Newcastle <56>. Thus, just as his father had not forced him to follow the trades he had first chosen for him, so A.E. did not force his son to follow his own trade. A.E. worked at Percy Main for about three years and "dwelled at Howden": since his second son and his dog joined him at one time, he must either have had a house of his own or have lived with a well-disposed lodging-house keeper <59,60>. The dog later was taken to "My Fauther's and there it was a gardien to him and Sister" - if not a slip, the singular may indicate that only one sister remained in the Felling house, the other(s) having left or died.

The autobiography has a curious and confused story to tell about this period at Percy Main. A.E. was a widower in his mid-thirties, and he tells us how he was nearly inveigled into marriage with a woman called Jane who later, in refuge at Tweedmouth, gave birth to a son which A.E. indignantly denies was his, adding that he had "had Nothing to do with the woman" - and, anyway, she turned out to be married already. Nevertheless A.E. had the child checked out by a family friend, who reported "it was none of Anty's Get, his Children had Strong bone in them"; and the story also contains a reference that raises the suspicion that a woman who may have been A.E.'s landlady, then or earlier, was in the plot <66>. Wherever the truth lay, this was undeniably A.E.'s second entanglement with a Jane.

In 1814 A.E. moved to Fawdon Colliery. After staying in lodgings for two months, he "got a house and coals at Hensharber" - free coals being a perquisite that went with a tied house - and his daughter, Ann, aged fourteen, joined him, to be his "houskeeper" <67,68>. But after a short time she caught a fever and died, "Which was a Loss greatly felt". A.E. then engaged "a houskeeper that I had before at Felling" - presumably she had either helped when his wife was ill, or been employed later at his father's house. He continued to visit Felling at weekends <**71**>. In 1816 or 1817 he moved to Gosforth Colliery, where "it was 3 weeks before I got a house", and he probably continued in this house when working for the same employer at nearby Coxlodge Colliery, after August 1817 <**74**>. But in September 1818 he moved again, to Backworth Colliery. Here his second son, Anthony Fenwick (his second name from his maternal grandfather), who was probably aged about 16, joined him in his work: he had been at school when A.E. was at Percy Main <**60**>, but, although unmentioned, may have begun to assist his father at Fawdon and Gosforth.

His second marriage

On 1 April 1818, A.E.'s father died, three days before he reached the age A.E. claims for him, 86. It was perhaps at this point that A.E.'s surviving unmarried sister(s) went to live with brother William. Although by now probably only the two youngest of A.E.'s five children still needed a home, A.E. knew where his duty lay. "I was then intent on Marrieing and Our Nibour said, 'Surely you will give Mary the forst offer?' " <**75**>. A.E. did, and married Mary Pearson, his second wife. Why she deserved the first offer was not revealed, but one possible explanation can be suggested. No record of the marriage has been traced so its exact date is not known. But it occurred after the death of A.E.'s father in April 1818, yet apparently before A.E. moved to Backworth in September of the same year. If it occurred after May, then Mary, like A.E.'s first wife, was pregnant at marriage, since a daughter was born in late February 1819.

There is, however, an alternative or supplementary explanation. A.E.'s laconic anecdote can be interpreted as indicating that Mary Pearson was already living with the Erringtons as a servant. She may have been working for A.E.'s father, in which case she lost her employment when the old man died and A.E.'s sister(s) moved. If she had helped to look after A.E.'s younger children, it made sense for him to make a home for both her and the children. But it is equally possible that she was the housekeeper originally engaged at Felling, who rejoined him in 1814 at Fawdon, and that she had been with him since. If this was the case, then all of A.E.'s children may have ceased to live with their grandfather some years before he died. Robert most likely lived in Newcastle after he was apprenticed in 1812/1813, and Anthony Fenwick may have worked and lived with A.E. from Percy Main onwards, taking the place of his brother. When, one morning in late 1815, A.E. walked to Heaton Colliery he "returned home to the Femely" <**70**>, which may indicate a family household at Fawdon instead of the previous one at Felling - even although he still visited Felling <**71**>, no doubt to see his father. And when A.E. proposed to Mary Pearson and she spent a night away consulting her mother <**75**>, then if the latter lived in Felling (a possibility indicated below), this suggests that the proposal took place at Gosforth/Coxlodge, and that A.E. was regularising an arrangement with a housekeeper, and apparent bedmate, who already looked after his two younger children.

'Mary Pearson' was a not uncommon name. The Newcastle Catholic chapel register records a Mary Pearson/Pierson of the Anglican parish of Jarrow who was born in 1804, and a Mary Pearson of Newcastle who was born in 1799 [CRS 1936, pp.221,231]. But although the parents of the former individual were acquaintances of the Erringtons (Mary's father was godfather in 1803 to a child of a local Catholic couple to whose later children A.E. and his sister Margaret were godparents [CRS 1936, pp.228,231,242,261,264]), we almost certainly need a Mary Pearson older than either of these. The 1851 census, taken three years after A.E.'s death, named a widow and "pauper", Mary Errington, born at Felling, who was living in High Street, Felling with her daughter, Mary Ann, also born at Felling [1851 census, enumerator's book for Heworth/Felling, PRO, HO/107/2401/ED5, entry 161]. This was the only Mary Errington at Felling, and the age of the daughter was given as 20, the correct age for A.E.'s daughter, Mary Ann. The age of Mary Errington was given as 60. Assuming that this was indeed A.E.'s widow and that she stated her age correctly, then this Mary Pearson should have been born in 1791. The Newcastle Catholic register has a gap between 1786 and May 1792, but a Mary Pearson of 'Jarrow' is recorded as a godmother in 1804 [CRS 1936, p.231]; however, it is doubtful whether a Mary Pearson born in 1791 would have been a godmother at the age of only thirteen. The Heworth Anglican baptism register records, as the only 'Mary Pearson' between 1785 and 1795, one Mary Penelope Pearson, daughter of John Pearson, baptised 15 April 1792. This Mary Pearson, who was most likely a non-Catholic, appears to have survived early childhood since she is not named in the Heworth burial register up to 1799. If this is indeed the right Mary Pearson of Felling, then she was in fact aged 59 in 1851, not 60. Whichever Mary Pearson became A.E.'s **second** wife, at marriage she must have been aged 26 or 27, and her **fourth** and probably last child was born when she was aged 40.

Presumably A.E., his new wife, his two daughters, now aged about twelve and eight, and the new daughter born in 1819, lived together in a tied house at Backworth. When A.E.'s natural daughter ("my doughter Sarah") was insulted by another girl, she went home "Crying and relating to Mother" <**79**>. A.E. makes no further mention of his second wife (unless a shopping episode relates to her, rather than to the first wife <**45**>).

His renewed domesticity did not prevent A.E. from **moving** again, to Walker Colliery; but at this point the autobiography **breaks** off. It contains clues, however, which suggest that A.E. may have moved to Walker after only a few years at Backworth. The move may be signalled within the account by the reference to a reconciliation with Straker, the viewer with whom A.E. had previously fallen out, a reconciliation which apparently led to the repairing by Straker of the Errington family fiddle, an event which A.E. dates to 1822 <**50**>. It may be significant tnat there are very few Backworth anecdotes; and one of these states that, only thirteen months after the episode which led to A.E. vowing to lay off drink, he was working at Walker and had already been made ill by the gas there <**79**>. This telescoping of the Backworth experience would fit a move to Walker in 1822, and if so, then A.E. most probably began his autobiography at Walker. But further moves may also be suspected.

We know from the Catholic registers that A.E. had four recorded children by his second wife [CRS 1936, pp.287,297; PRO, North Shields Catholic register; St Andrew's chapel register, 2:42]. The fact that the baptism of a son in May 1825 was recorded in the register of the

North Shields chapel makes it likely that A.E. was living in the vicinity at this date - perhaps, after all, he had not yet moved from Backworth, or alternatively, he had returned there. The latest episode recorded in the autobiography dates to 1827 and occurs when A.E. is visiting South Shields <**69**>. It is arguable that he is less likely to have made this visit if he was living north of the Tyne than if he was living again at Felling. His last recorded child was in fact born at Felling, in 1831 (and the baptism took place at Newcastle). Since A.E. was only aged 54 in 1831, it is likely that he was still working, and plausible that he was again employed at Felling Colliery, where he may have been since 1827 or a little earlier.

A.E. died at High Felling on 31 January 1848, aged 69 years, his occupation being registered as 'waggonwright'. (The registrar at Heworth was George Sill, perhaps a descendant of the 'Mr Sill' to whom A.E.'s first wife had been housekeeper before her marriage <**31**>.) A.E.'s youngest child, his daughter Mary Ann, was the witness named on his death certificate, but as we have seen, his wife outlived him. In 1851, A.E.'s widow and daughter were living in a house two tenements up from the corner site on the west side of Felling High Street which shortly afterwards became the 'Beeswing' public house (PRO, **HO/107**/2401/ED5); and it is plausible that this was where A.E. died three years earlier. At the date of the next census, 1861, Mary Errington was no longer living in this house and was probably dead.

A family man

During a period of over thirty years A.E. was supporting the children of his two marriages, although for about one quarter of the period he was a widower. Before his first wife died he may have lived in the same house for some ten years, but thereafter, while his children grew up in the security of their grandfather's home, A.E. moved his weekday residence many times, probably more than seven times. (Migration from colliery to colliery was common among Tyneside colliers of the period [Gill and Burke 1987], but whether colliery tradesmen moved as frequently is not known, hence it is uncertain how typical or untypical A.E.'s behaviour now was.) He lived sometimes in lodgings, but preferably only for short periods, and latterly he seems to have had his own house and to have shared it with a child and/or a housekeeper. His movements were confined to Tyneside, when he might alternatively have sought employment further away, on the Wear: although there may have been other reasons, such as his contacts with Tyneside employers, it is likely that a major reason was his desire to be within weekend visiting distance of his children at Felling. He does not record all their names or note their births individually, but patently he had a strong attachment to them, and the respect and affection he felt for his own parents he clearly expected to be generated between his children and himself. Anecdotes express a degree of pride in his sons and some concern for the health of one of them; and his affection for his natural daughter, Sarah, was reflected in his shame when she was told that her father was a drunkard, and in his consequent vow to give up drink <79>. Although we only have A.E.'s own view of his family life, probably it was, as he indubitably thought it was, a satisfactory and happy one. A.E. was a family man and, despite the deaths of his first wife and a daughter, a man who enjoyed his role as a family man.

APPENDIX

(a) the family tree (b) local Erringtons (c) possible North Shields siblings of A.E. (d) A.E.'s pregnant brides

(a) **THE ERRINGTON FAMILY**

William
farmer, Netherwitton
m. 2.6.1725* **Mary Avery** (Netherwitton)

Robert 4.4.1732* (Netherwitton) - 1.4.1818* (Felling)
w-wright, Felling ?1754-1799
m. 8.12.1766* **Isabella Carr** of Matfen 1742 - 20.4.1802* (Felling)

William	**John**	**George**	**ANT.**	**Isabel**	**Jane**	+ 3 unnamed daughters
10.8.72* - ? ; c.1820 shipman, North Shields; ? 1818° child bpt	4.9.74* - ? ; 1792 w-wrt, Felling; 1825° godf John	6.10.76* - 19.6. 1834* soldier		1780 - 19.1. 1791*	? - ? 1806 godm Isabel	? Margaret 1825° godm John; ? Sarah, ? bro-in-law's wife 1792° Sarah Clarke nee Errington child bpt; ?

ANTHONY

20.9.1778* (Felling) - 30.1.1848* (Felling)
w-wright, Felling 1792-1809, then elsewhere
m. (1) 28.5.1798* **Ann Hindmarsh,** d. of Fenwick Hindmarsh, 15.5.1768* (Kkwhelptn) - 9.3.1809* (Felling)

Robert	**Ann**	**Anthony Fenwick**	**Isabel(la)**
29.10.(1798)* - ? ; 1812-1819 painter Newcastle	1800-1814 (Fawdon)	? - ? ; 1818 w-wright Backworth	21.9.1806° - ? ; 1820° godm, North Shields

<continued on next page>

COMMENTARIES : FAMILY

ANTHONY

m. (2) 1818 **Mary Pearson**
? 15.4.1792 (Felling) - ?

Mary	**George**	**John**	**Mary Ann(e)**
29.2.1819°	6.4.1821°	2.4.1825°	12.9.1831°
bpt North Shields	bpt N-castle	bpt North Shields	bpt N-castle
- ?	- ?	- ?	- ? ; 1848 witness at father's death

and (3) by **Jane Richardson**

Sarah
12.8.1810° (Heworth), bpt N-castle, - ?; 1831° godm to Mary Anne

KEY

* = the Anglican parish registers of Netherwitton (for A.E.'s father's birth and his grandfather's marriage); Kirkwhelpington (for A.E.'s first wife's baptism); South Shields (for the death of A.E.'s brother George); Heworth chapelry (for all other events). The latter register records, separately, the births - not baptisms - of four sons of Robert and the oldest son of Anthony Errington, but in the case of Anthony's son, Robert, the year-date is not stated.

° = the Catholic registers of Tynemouth and North Shields chapel (for the birth of John Errington) and of St Andrew's chapel, Newcastle (for all other events), that is, CRS 1936, up to 1825, thereafter the manuscript register. Note that the printed register covers baptisms 1765-1825 and rites of last unction 1765-1806, but the baptisms 1774-1792 are very incomplete and the remaining baptisms probably slightly less than complete. Hence the deficiencies of the register may in part explain the absence of records for some Erringtons. Note also that the Catholic registers record birth-date as well as baptism-date, and up to c.1815 the Anglican parish of residence, but not the occupation of the father. The dates cited in the family tree are birth dates.

Other dated events are evidenced by the autobiography.

Notes on the family tree

(i) **Catholic registers** A Jesuit mission (strictly 'ex-Jesuit' up to 1803 because of the formal dissolution of the order) existed in Newcastle up to 1807, apparently with a public chapel, but its records are lost. St Andrew's chapel in Newcastle, served by secular clergy, probably had the larger congregation, and after 1807 it was the only Roman Catholic place of worship for Newcastle, Gateshead, and the settlements on both banks of the Tyne down to the sea. In 1817, however, a chapel was built at North Shields for the lower district, and this eventually kept its own records. The only Erringtons recorded in the printed register of St Andrew's appear all to have been members of A.E.'s family; but the failure of this register to record the baptisms of any Errington before 1806 may indicate that the family went elsewhere for baptisms or used the services of a priest from elsewhere. The territorial impermanence of pre-1830 Catholic chapels, priests and registers makes it difficult to know in which Catholic registers a Catholic family's baptisms and deaths would have been registered - if indeed they were registered, since it is clear that this did not always occur. Up to the date of secular registration, 1837, I can find no reference to Robert or Anthony Errington or to any of their families in the registers of those Catholic chapels within reasonable distance of Felling, other than the chapels at Newcastle and North Shields. That is, there is no reference in the baptismal registers of Stella (county Durham), Cheeseburn Grange (in Stamfordham parish in Northumberland, but including some Stella entries), Pontop/Lanchester (Durham), and the two chapels at Durham (the Jesuit chapel register has inserted a few entries from Felling 1837-1841); or in the death entries in the Pontop register [PRO, DCRO]. This strengthens the possibility that up to 1807 the Erringtons were recorded in the non-extant registers of the 'Jesuit' chapel at Newcastle.

(ii) **William Errington, A.E.'s grandfather** Although A.E. says that his grandfather retired "being aged", and thereafter his son became a waggonway wright for 45 years and retired in 1799, in the 1750s the grandfather had only been married thirty years (since 1725) and is therefore unlikely to have been much over 60. The first page of the Heworth register, recording the births of Robert Errington's sons, erroneously describes the father as 'William', the name then being partly erased. This suggests that grandfather William was still alive in the late 1770s, was living with his son at Felling, and was known to the parish clerk, hence the confusion over the name. If so, he would probably be in his 80s - but his son lived to be 86.

(iii) **William Errington, A.E.'s brother** The only entry in the printed register of St Andrew's for a William Errington records the baptism in July 1818 of a child, Anna Curry, to William Errington and Margaret Curry, of All Saints' parish, Newcastle - there may be an error in the register but the entry as it stands indicates a bastard. A.E.'s brother, William, was then aged 46, and appears to have lived at North Shields c.1820, points which perhaps raise doubts about the identification.

(iv) **Margaret Errington, A.E.'s sister** The evidence that one of the three sisters of A.E. who are unnamed in his account was called Margaret is as follows. Margaret Errington appears in the printed register of St Andrew's chapel as a godmother many times between 1811 and 1820, including four times in relation to families specified as being of the parish of Jarrow/Heworth. In 1811 and 1813 she was godmother to children of William and Elizabeth Thirlwall/Thelwall of the parish of Jarrow/Heworth; and in 1807 A.E. was godfather to an earlier child of the same couple.

(v) **Mary (Pearson) Errington** A.E.'s second marriage in 1818 is not recorded in the Heworth marriage register and Mary Errington's death is not recorded in the Heworth burial register up to 1866.

(b) The local Erringtons

'Errington' is a fairly common name in North East England. Various Erringtons, and even more than one Anthony Errington, were burgesses of Newcastle from at least the sixteenth century onwards, while at one time an Errington was the town fool. Erringtons also existed in the countryside. The Erringtons of Beaufront, although continuing as Roman Catholics, were represented in the later eighteenth century by an individual who was both an extensive landowner and Provincial Grand Master of the regional Freemasons; and these Erringtons assisted in the building of St Andrew's chapel at Newcastle, where A.E. worshipped [Sykes 1833, 1:308]. A priest at this chapel in A.E.'s lifetime, George Errington, later rose to be an archbishop [Vincent Smith 1930, p.78]. An earlier Anthony Errington was a Catholic writer of the early eighteenth century [DNB]. Contemporary with A.E. were four local Erringtons who in 1825 subscribed to Mackenzie's 'History of Newcastle' [Mackenzie 1827], and the Erringtons in Gateshead included a skinner and "the first beer-house keeper under the new act" [Mackenzie 1834, p.101]. While most of the Erringtons in the Heworth parish register and the Newcastle Catholic register during the period of the autobiography were members of A.E.'s family, the number of Tyneside Erringtons appearing in the public indexes of the national secular registers after 1837 is so considerable as to make tracing the surviving members of A.E.'s family impractical.

(c) Possible siblings of A.E. in North Shields

Directories of North Shields record a Margaret Errington, of 16 West Percy Street, who in 1822, 1827 and 1834 ran a "ladies' academy", i.e. a girls' school, and a William Errington, shipmaster, who at the latter date lived next door, at 15 West Percy Street [Directories 1822,1827,1834]. Were they A.E.'s sister and brother? If so, both had risen higher in the world than A.E. William's promotion seems plausible, since the autobiography suggests that he had some connections with shipping and perhaps with a specific ship [**81**] - Margaret's less so, given A.E.'s limited schooling. Moreover, if the

Margaret Errington of the "ladies' academy" was the same as the "schoolmistress" Margaret Errington of the 1841 census, allowing that she had moved since 1834 from West Percy Street, where there were no longer any Erringtons, to Norfolk Street, her age, now stated, was only 40. This would make it impossible for her to be A.E.'s sister, since her birth would have occurred when his mother was in her late 50s. However, since Margaret of Norfolk Street appears to have been the head of a household where another member, Winifred Errington, was aged 50, it is just possible that Margaret's age has been misread and that it was actually 60. It is perhaps more plausible that she ran a 'ladies' academy' at age 41 (in 1822) than at age 21, and if so, there is still an outside chance that she was A.E.'s sister. There were, however, other Errington families in North Shields and Tynemouth.

(d) **A.E.'s pregnant brides**

It looks as if both of A.E.'s wives were pregnant at marriage, a circumstance common in the general population for centuries. The Roman Catholic registers do not lend themselves to investigation of the historical frequency of bridal pregnancy and I know of only one result, that reported by myself [Hair 1968]. A small sample from a single eighteenth-century Lancashire register suggested that the Catholic frequency was much the same in that period as that of the nation at large.

COMMENTARIES

A.E.'S RELIGION AND MORAL VALUES

His religious affiliation

A.E. was a Roman Catholic. But this aspect of his life is only lightly touched on in his autobiography - although it may have been fundamental to its production. Only one anecdote makes his religious affiliation fully explicit. A woman he has assisted remarks that she had not thought that a papist would behave so to a protestant - she having previously refused to do A.E.'s father a good turn on the grounds that he was a papist <54>. Two other anecdotes refer obliquely to A.E.'s denominational allegiance. An opinion on ghosts is provided by a "Cathlick Preast, Newcastle upon Tyne" <31>, and a vision is vouchsafed to A.E. when he walks home with a friend expressly stated to be a Catholic <**70**>. In a handful of anecdotes, A.E.'s behaviour almost certainly bears some relationship to his being a Catholic, for instance, the anecdotes concerning his providential capacity to make butter set <**20-22**>. But, in contrast, in many anecdotes A.E. uses the language of contemporary mainstream Christian ideals and devotion, but not those terms which might distinguish papist from protestant. Finally, it is very difficult to be sure whether or not the quirks of behaviour documented in many anecdotes have anything to do with his religion; and equally difficult to be sure whether or not his simple saga of events can be stripped down to reveal an underlying Catholic network of acquaintances and influences. However, the extreme positions - either overstating the influence of A.E.'s Roman Catholicism, or undervaluing it - can perhaps be avoided when the Catholicism of the period is considered.

During the decades covered in the autobiography, the English Catholic community, a tiny minority in the nation, while recognising that it was separated from the Church of England, was less 'set apart' mentally and socially than it had been during the earlier centuries of penal segregation, or than it was to be after the nineteenth century 'Second Spring'. This is the view of recent Catholic historians who emphasise the extent to which the English Catholic laity, partly in response to eighteenth-century Anglican tolerance, felt themselves to be within the mainstream of Christian devotion and piety, in their contemporary English forms. Indeed, one historian applauds this period of modus vivendi as a foreshadowing of post-1960s Catholic ecumenicism. In his view the 'Second Spring' was a take-over, by the Irish, the clerical hierarchy and Rome, of an indigenous Catholic tradition in England not itself unworthy; and the ending of the modus vivendi is deplored in a chapter entitled 'The Victory over the Laity' <Aveling 1976>. More cautiously, another historian nevertheless notes that the eighteenth century Catholic community tended "to share the essential secular values of the society it lived in", to practise its more assertive papist beliefs only in the region of private family life, and in general to accept that it was only one Christian denomination among many <Bossy 1975, pp.110,284,386>. A.E.'s apparent reticence about his Roman Catholic allegiance is in line with contemporary Catholic conformity to a generally accepted modus vivendi; and as with all such conformity, it is perhaps unwise for the historian to probe much further, in search of individual motivation.

A.E. begins his autobiography with expressions of Christian belief. These include a belief in divine providence, and his mainstream, common-ground Christianity is suggested by the fact that he cites for this purpose an anonymous author who was actually the Anglican evangelical, Hannah More (at least one of whose many contemporary tracts he seems to have read) <1>. Next, his father is commended for being industrious, honest, well-liked and a healer of discord - albeit the last quality is probably to be linked with his love "to be in Sochiel Company" <2>, which we can probably interpret as a reference to public drinking. When he narrowly escapes falling down the pit shaft, A.E.'s father kneels and thanks God; and A.E.'s filial respect is such that he does not appear to notice that on one underground occasion the old man acted rather incautiously <2,14>. His mother, A.E. says, was industrious and dutiful. More specifically Christian values are only broadly invoked. "His Law was to us all to be honest, to be Charatable, to shun Bad Company, and to keep the Commandments in a Christin life. And to love each other was the Charge from our parents". But A.E. concludes these opening passages with a simple Christian prayer <2>.

His moral values

A.E. seems to have been a fairly normal boy: he dodged school, slung stones with his garters, and was scatalogically-minded <**4-6**>. He appears to admit these naughtinesses, and there are some childhood episodes which are not particularly didactic - foreshadowing the Pip of 'Great Expectations', he reads the gravestones in the churchyard (a sister of his had probably just been buried there); he fights another boy successfully; he takes a dip in the river and learns to swim; he is pleased when bullies are caned <**4,5,10**>. He makes no special miraculous claim for recovering his sight in one eye, he is only mildly aggrieved when flogged unfairly by his schoolmistress, he saves his own life in the river by commonsense <**7,4,10**>. He utters patriotic sentiments, not altogether unreasonably, given the wartime context <**23**>. But also he was already becoming providentially ordained to do good works, finding poisoned ducks and a pistol, and reporting the same, or fetching a taylor's goose, and already he enjoys being praised for his various actions <**7-8,25**>. His miraculous intervention in butter-setting we will return to <20-22>. His Sunday meeting with Tom Bilton would be a simple adventure if he did not introduce the term "Shaddow" <**26**> which seems to have had faint supernatural overtones for A.E. - "This was the Shaddow taken for the real man" <**41**>. He already speaks his mind (or this is how he recalls it), complaining about jobs he does not like, and addressing army officers <**13,15,23**>.

As a youth, A.E. saves the lives of men in the pit shaft and on the colliery waggonway, but only by quick thinking, and no cosmic moral is drawn <**11-12**>. He is publicly commended for his skill in wood-working, but only shows natural pleasure <**18**>. The anecdote about the missing horse is without any pretentious claim and is merely a heart-warming story, not least for animal-lovers <**24**>. The tone becomes more mystical and didactic when an older collier tells him about a vision of disaster he has had and makes A.E. listen to him praying - A.E. "remained silent", perhaps because he had doubts about being associated with a protestant in prayer, but he kisses the Book, presumably the Bible <**19**>. When he saves the Manchester waggon, he refuses to accept money for acting on behalf of God's Providence (but his father accepts it, more realistically), and is described as "a wonderful Sun" <**27**>. Twice saved in the shaft, he thanks Our God for deliverance. But as with many later anecdotes invoking divine providence (which it would be tedious to cite in detail), not least those relating to underground dangers, the moral of the escape from peril is also - inevitably if not intentionally - self-congratulatory, A.E. having always displayed quick thinking <**28,32-33,36,40,47,48,51,52,57,62,66,73**>.

Providence saves many individuals by A.E.'s intervention, and it saves A.E. himself - from an unsuitable second wife. Here and there the element of chance which brings A.E to the scene of potential disaster is indeed loosely 'providential'. As a Christian A.E. is entitled to believe that, in some sense, all human actions are providential. But the notion is patently being overworked, morally speaking, when A.E. is providentially led by the Divine Spirit merely to be a sightseer at the scene of a disaster, or more distastefully, when he intervenes in the love-life of his Benwell landlady and then is "driven in Speret" to drop in on the subsequent family row <**48,70**>.

A.E. reports that he is many times thanked, often in glowing terms, for his dramatic actions, some of which, admittedly and mercifully, he forgets to tell us are providentially determined. If the episodes are all fully truthful or accurately told, it appears to have been often the case that it was his knowledge and experience, particularly of underground conditions, as much as, perhaps more than, his quick-wittedness, that saved the day. While the providentiality of certain of the episodes seems very thin indeed, in the sense that A.E. appears to drag in Providence only as a moral and rhetorical signing off, it need not be doubted that his overall belief in the support of divine favour was very strong or that his sense of providential direction was normally sincere. In fairness, it must also be added that some anecdotes are free from either providence or self-congratulation, and that a few even depict A.E. unfavorably - "agains myself" as he promises in his introduction <**1,38**>. In a single instance, he almost joins in the laughter of others at his own superstitious fears <**81**>.

A.E. often represents himself as acting from common kindness, and no doubt this was so. He is kind to a hungry milkmaid, and he is "touched with compassion for the poor and needy" <**69,63,64**>. His responses are practical, and thus in public houses he collects money for beggars, quite substantial sums from a small number of givers, indicating that A.E.'s sympathy was shared among fellow workmen - and also that his approach to others was sometimes taken more seriously, and was less intrusive and counter-productive, than a critical reading of the autobiography might lead one to suspect was at other times the case. The Burns impostor also receives generous treatment in a public house, possibly at A.E.'s initiative <**29**>. Other acts of kindness include helping a hard-pressed shopkeeper <**51**>, helping a drunken woman home <**65**>, and rescuing those in difficulty on the roads <**40,54,73**>.

As is now generally recognised by historians, Victorian 'decency', alias 'prudery', preceded the Victorians. A.E., a working man writing in the 1820s, is very careful to use polite language, even when describing events unsuitable for drawing room conversation, and even although so many of the episodes occur in public houses where no doubt the language often became repetitively impolite and indecent. He refers to "excrement" and "made water" and he goes to the "necessary" <**6,35,52**>. He allows "arse" once <**45**>, but abbreviates a swear-word and writes "B..... your eyes", "B..... the dog", and "you Bu....." <**41,59,60**>. He notes that "Bad Language" is used by a drunken man but does not repeat it <**54**>. A pubic hair is from "you ges from where" <**35**>, and this is the nearest A.E. gets to bawdy. Two anecdotes are scatological but decently expressed <**6,38**>. Expletives are mild - "Good God!" or "God!" <**22,43,51,57,73**>, and once "D..., there its!" <**62**>. A man says "Damn the Devell to hell", but this is to demonstrate that he is "an unbelever of Athiest Cast of mind" <**44**>. A deceived woman utters "Satan, begone!", A.E. apostrophizes a donkey as "Poor sillie thing" <**81**> and after a quarrel with his bosses he gives himself only the dignified exit line, "Good day, Jentlemen" <**76**>.

A.E. only once mentions other denominations, and never explicitly criticises their beliefs - although he cannot have avoided being aware that he was surrounded not only by a protestant community which, when mildly aroused, could be dismissive and derisive towards papistry, but by a minority of sincere protestant activists. The single reference is to the Baptist preacher who adulterously and bigamously courts A.E.'s landlady for her money. A.E. also notes that he "saw him give her 18s 0d which he said he had got that day for preachen at Newcastle" <**48**>. The preacher may have been merely paying his board, but A.E. clearly implies that he only preached for mercenary reasons and was a hypocrite. The nearest A.E. gets to doctrinal criticism is when he implies that two men arguing about a Bible text - A.E. is capable of quoting the Bible himself but lay arguing surely indicates that these are protestants - are not permitted to share a vision on the Town Moor <**70**>. Contrariwise, A.E. is perhaps recording evidence of some degree of positive tolerance towards other beliefs when he remarks, albeit a little obscurely, that concerning the Jews he "had the Contemplation that they would bee up on the Day of resorection as well as those in the Church yard" <**81**>.

Industry, job-satisfaction, family loyalty, patriotism, quick-wittedness, speaking one's mind, general kindness, verbal decency and propriety, and cautious tolerance - these are not specifically religious and Christian values, just as an awareness of divine providence is not specifically a Catholic value. What then were the manifestations of A.E.'s religious affiliation ?

The Erringtons as Roman Catholics

A.E.'s father was born at Netherwitton, a Northumbrian village characterized in the 1820s as follows: "some years ago the inhabitants were almost entirely of the Roman Catholic persuasion" [Mackenzie 1827:2,155]. (This Catholic community probably resembled the 'seigneurial' Catholic communities further north in Northumberland recently studied in detail [Bossy 1967-8, 1969-70].) The parish register of Netherwitton chapelry records Robert Errington's birth in 1732, not his baptism - at a later date the register kept a completely separate record of the births of 'dissenters' [Steel 1974, p.854]. Presumably Robert was baptised by a Catholic priest, but no record has been traced. When Robert's four sons were born in the 1770s at Felling, a separate page of the register of Heworth chapelry (the first page of a new volume) recorded their births, apparently on Robert's specific request, since no other Catholic births were noted. It is highly likely that the sons, and also their five sisters, were baptised by a Catholic priest, perhaps at Newcastle, but the records of the two chapels there are lacking and defective, respectively, in relation to the period, and no record of these baptisms has been traced. (Alternatively, they may have been baptised at home by an itinerant priest from a more distant chapel, and recorded - if recorded - in his or the chapel's register. It needs to be borne in mind that not only are some contemporary Catholic registers not extant, but those that are extant are often defective and generally provide less than comprehensive coverage for their wide areas.)

By the mid 1760s, and perhaps as early as the later 1750s, Robert Errington was living at Felling and working for the Brandlings. In these decades, the Brandlings were Catholic; they had a private chapel at Felling Hall until at least 1756; and there was a Catholic community in the Felling district estimated to run into several hundreds, possibly the largest on Tyneside [Walsh and Forster, pp.53-5; Bossy 1975, p.301]. It is difficult not to believe that Robert Errington's removal to Felling had something to do with the Roman Catholicism of the Brandlings - although this link evaporated in the 1770s, when the head of the family renounced Catholicism. When in turn A.E. came to beget children, of the four born between 1799 and 1806 to his first wife, the birth of the first, a son, was recorded in the Heworth register, below the entry recording the birth of A.E., but the register contains no reference to the other three children. However, the baptism of one of them by a Catholic priest at a Newcastle chapel is recorded in the chapel register, and perhaps the other two were baptised, if not privately at home, then at the other Newcastle chapel, the one whose records are lost. The four known children of A.E.'s second marriage, as well as his illegitimate daughter, were all baptised in the Catholic chapels of Newcastle and North Shields. No records of either their baptisms or births have been traced in Anglican registers, and this was no doubt because by these later dates it was neither legally nor socially necessary for Catholics to make use of the Anglican registration system.

As the local Catholic registers show clearly, papists frequently married protestants. It is just possible that A.E.'s mother was, before marriage, a protestant, but more likely that she was not. (Her birth/baptism is not recorded in Stamfordham parish register, but has also not been traced in any Catholic register). However, A.E.'s first wife was almost certainly a protestant. (She was baptised at Kirkwhelpington church and no record relating to her or to her family has been traced in Catholic registers - for instance, no individual called Hindmarsh appears in the Capheaton/ Kirkwhelpington Catholic register [CRS 1914]). A.E.'s second wife cannot be identified with absolute certainty but there is a strong likelihood that she too was a protestant. As the law required after 1753, both A.E.'s father and A.E. himself participated in marriage ceremonies in church, at Heworth chapel. Catholics were supposed to be instructed by their priests to participate in the Anglican marriage service only minimally (by turning their back on the clergyman or walking about the church during the service). But it is generally doubted whether this degree of protest was normally followed, and it is even uncertain whether there was common acceptance of the more positive procedure of a separate marriage ceremony, performed before or after the Anglican one, by a Catholic priest, usually privately [Steel 1974, pp.863-7]. There is no record of Catholic ceremonies of marriage involving either of the Erringtons, but this may again be due to gaps in the Catholic registers. A.E.'s first wife and probably also his second wife were pregnant at marriage, but this degree of laxity was so common in protestant England, and so unpunished by the later eighteenth century, that it may be doubted whether it was regarded, equally among Catholics, as more than the most trifling moral lapse. Probably little significance should be attached to the fact that A.E. does not disclose these lapses.

In this period Catholics were normally buried in Anglican churchyards, and Catholic registers do not normally record deaths or burials [Steel 1974, pp.885-6]. The Catholics of Tyneside occasionally made use of the dissenters' burialground in the Ballast Hills [Steel 1974, p.891], but the Margaret Errington who was buried there in 1799 cannot have been a member of our family of Erringtons [Miscellanea 1930, p.264]. A.E.'s parents, sister, and first wife were buried at Heworth, as he himself was to be.

Because there are many gaps in the record, our knowledge of the attitude of the Erringtons to strict adherence to 'ultra' Roman Catholic practice with regard to the rites of passage is to some extent uncertain. But what we do know makes it look as if they conformed to current conventions in the English Catholic community, that is, to an accepting modus vivendi. About the private devotional life of the family, the area in which Roman Catholic historians believe that specific Catholic practice had most sway, A.E. tells us nothing. To Catholic public worship there is only one reference, and that an unhelpful one: A.E. walks several miles on a Sunday to attend the Catholic chapel, but is distracted by meeting another boy and dodges the service <26>. There is, however, some evidence that A.E., as an adult, was active within a self-consciously introvert Catholic community. The evidence is provided by the Catholic registers, which show A.E., his siblings, and his children, appearing fairly regularly as sponsors, that is, god-parents, at baptisms. Not only does A.E. contribute to the network of Catholic inter-relationships, but it is possible to identify a tighter network of A.E. and his Catholic friends (or at least close acquaintances), by listing the families for whom A.E. acted as godfather and the families who provided godparents for his children. Finally, there is the curious fact that A.E. had his illegitimate daughter Catholicly baptised and therefore recorded in a register which has almost no other bastards - one interpretation of which might be that A.E., unlike his fellow Catholics locally, was determined to purge his shame by regularising and to some extent publicising his slip.

If the extent of A.E.'s inner Catholic piety is difficult to assess, so is the extent to which A.E. operated within a Catholic network. It seems likely that his schoolmistress, who was Irish, was a Roman Catholic; but his schoolmaster almost certainly was not, for Heworth school must have had a church connection, since Mr Glover, the incumbent, is mentioned in connection with it <5,9>. Mr Glover had married A.E. and his first wife, and recorded the birth of their first child, and later incumbents of Heworth officiated, at least notionally, at the burials of A.E.'s relatives. (One of the incumbents, John Hodgson, played a major role in the relief of the 1812 Felling disaster, and also in the discussion about safety in mines which followed and in which A.E. was to play some small part himself; but by 1812 A.E. had moved from Felling and presumably had no occasion for personal acquaintance with Hodgson.) When a child, A.E. had played in Heworth churchyard, and churchyards of that period regularly contained graves of Catholics - St. Nicholas' churchyard in Newcastle, which A.E. mentions <**52**>, contained the grave of at least one Catholic priest [Mackenzie 1825, p.365]. However, A.E. only specifies one family burial, that of his daughter at Gosforth <**68**> (a burial which curiously cannot be traced in the Gosforth register). But he mentions several non-family burials in Heworth churchyard: as a child of eight he acts as a pall-bearer at the funeral of his school-mistress, and as an adult, when men are killed in the pit, he is active at the burial - "I Lowered Riddly into the grave" <**43**>. A.E. knew enough of the Anglican parish system to speak of "the Cheplry of Hueth, in the parish of Jarrah", a detail he probably was taught at his second school. But whether these tenuous Anglican connections meant anything to him is unclear.

Around 1820 a Brandling spoke warmly of "my old servants the Erringtons", and A.E. worked for the Brandlings at Gosforth after he left Felling <**74**>. But a little later a Brandling did not support A.E. against his viewer, causing A.E. to walk out <**76**> - and even if the original link between the Erringtons and the Brandlings was religious, in A.E.'s lifetime the Brandlings were protestant. As a result of the earlier Brandling allegiance, probably many of the original workmen at Felling Colliery were Catholic, but hardly any of the workmen whom A.E. names there (or at his later collieries, on the supposition of a Catholic helping-hand network), can be traced in the Catholic registers. Instead, many of the named Felling workmates can be traced in the Anglican register, not only in the burial register but also in the marriage and baptism registers, suggesting that they were not Catholics.

Religion and superstition

There remain to be discussed a dozen anecdotes in which A.E. recounts paranormal, 'supernatural', occult episodes, or expresses superstitious beliefs. Tynemouth Castle has an "enchantment" <**80**>. But there is evidence from a diary of 1780 that, in relation to the underground passageway known as 'Jingling Man's Hole' (nowadays

'Jingling Geordie's Hole'), A.E. was only repeating an earlier, popular, local belief [NSCL, Taggart typescript c.1930, p.31, the source for the diary reference unstated]. The pamphlet A.E. cites has not been traced, but a notice in a local newspaper in 1819 referred, albeit jocularly, to "the Whole Enchanted Secret of Jingling Man's Hole" [NSCL, Taggart typescript, p.32]. But other episodes show A.E. as a convinced and eager, indeed credulous, believer in the supernatural. When a youth, A.E. was told by an older man, apparently a non-Catholic, that he had had a vision of future disaster at Felling, from which, however, A.E. would himself escape. Writing several decades later, A.E. recounted the episode and clearly inferred that the prophecy had been fulfilled by his leaving Felling only one year before the 1812 disaster <**19**>. Since A.E. had worked at Felling for nearly twenty years before he left, and since he would almost certainly have been killed if he had stayed, his interpretation of his departure as providential is understandable. At one of the Felling pits, when an old man saw a ghostly "Spiret" - the premonitory figure of a man who was about to die - A.E., after being told, saw the ghost himself <**34**>. Much later, A.E. and a companion twice saw on the Town Moor a ghostly funeral procession, once as a premonition of the Heaton colliery disaster of 1815, which occurred more or less under the Town Moor, and once after the bodies had been recovered <**70,71**>. Given the scale of these two disasters, and in the case of the Heaton disaster the local myth that some men remained alive for months, trapped underground (a myth surprisingly accepted as fact in a standard history [Ashton and Sykes 1929, p.41]), and hence the over-wrought feelings of those in the locality who continued to work undergound, some degree of hysteria giving rise to superstition might be thought not unexpected. But, in fact, there is no evidence that the miners in general allowed play to any overwrought feelings they may have had, not even in the direction of experiencing or reporting supernatural happenings. (The belief in the survival of some Heaton men underground was wishful thinking, but in terms of the actual disaster, a flooding, not entirely irrational, at least for the first week or so.) It would be strange if, in the period 1790-1820 when large disasters and minor fatal accidents were so common in the North East that perhaps one in ten of the coalminers died from occupational violence each decade [Hair 1968, p.554], there was no talk of omens and premonitions. But again, so little of this reaction appears in extant records that it is difficult to believe that it was at all common. Against this background, A.E.'s own response - his invoking of the supernatural to balance the contingent - seems eccentric and individual.

Another ghost is seen by A.E. after a mining death, but in this case "the sole could not rest because there was money hidden" <**43**> - probably A.E. wants us to think, not so much that the man was a miser, as that he was anxious that his widow should receive the money. Less easy to interpret, except perhaps in tortuous Freudian terms, shortly after his marriage to an older, protestant woman, A.E. is persuaded by her that there is a ghost in the house, "a tall slender Women dressed in scie blue silk". Although A.E. sensibly does not immediately tell his wife that he too saw the ghost, the couple promptly move house <**31**>. On three occasions A.E. is pursued across open ground by fire-balls or lightning, once after consorting with "an unbelever of Athiest Cast of mind" who had challenged the Devil, once when a "wise man" (perhaps a traditional 'cunning man' [Thomas 1971, chap.8]) jestingly tested his courage, and once for no particular reason (unless it was for entertaining a kind thought about Jews) <**44,53,81**>. (It is uncertain whether the reference to the Freemasons' Lodge in the first of these episodes is significant in the present context. In the 1770s, a Roman Catholic Errington, a landowner, was Provincial Grand Master of the local Freemasons [Sykes 1833, 1:308], but by the 1800s when the episode occurred, or by the 1820s when A.E. recollected it, the Freemasons may have been linked in A.E.'s mind with godlessness.) A.E. also believed in spells - a nightwatchman put a spell on a thief, his qualifications as a magic worker being that "he was scield in Astronemy and understood the Power of the planets" <**30**>. Finally, A.E. himself, when a child, thrice makes butter set, once after an old woman puts a curse on the operation (not the first reported instance of an old woman, refused buttermilk, responding so [Thomas 1971, p.556].) Since on the first occasion he has to be instructed by his mother, it is perhaps unlikely that A.E. had a previous reputation as a magic worker in this direction <**20-22**>.

What connection, if any, does A.E.'s easy acceptance of the

paranormal have with his Catholicism? In general, it is conceivable that the greater emphasis laid within Roman Catholic practice on procedures involving sacramental beliefs - at the mass, at baptisms, in holy water and priestly blessings - procedures which protestants either did not follow or tended to interpret more in symbolic than supernatural terms, made papists more subject to credulity in relation to supposed supernatural happenings than were protestants. A recent Catholic historian does not hesitate to refer to the folk-religious character of the practices of the labouring class of Catholics, calling them a "complex of observances and beliefs, often superstitious", although admittedly in relation to the previous century [Aveling 1976, p.300]. Be that as it may, as regards A.E.'s Catholic contemporaries on Tyneside, we do not have the evidence (or at least no relevant evidence has reached me) to indicate that A.E.'s level of superstitious interpretation was at all typical of individuals of his social order within the Catholic community. Indeed, we have little sound evidence about the general tone of the corresponding protestant community, and it will have been noted that A.E.'s comrades in superstitious belief appear to have been sometimes non-Catholics. It is true that the view was sometimes expressed by post-Reformation writers that papists were inevitably "more superstitous and credulous ... especially the vulgar and unlearned sort of them" (Selden), but the trend of recent research has been to stress the extent of superstition among the protestant peasantry and to question a unique Catholic susceptibility [Thomas 1971, pp.273,408, and passim]. It appears that A.E. was first introduced to ghosts in the mine underground by a non-Catholic <**34**>, and many years later, when strange sounds were heard underground at Percy Main, the viewer arrived "with Bible and Prayer Book" <**62**> - that is, a non-Catholic was preparing for some form of exorcism

Nevertheless, specific links between A.E.'s religion and his superstitious beliefs can be found. A.E. is often eager (or so he says) to keep knowledge of occult manifestations secret at the time. When his wife sees a ghost, he tells her she is "not to menchon it to the nibours" <**31**>. An explanation is perhaps provided by the fact that later A.E. does not circulate the vision of the forthcoming Heaton disaster (which it might be thought would have been kinder to have done), and when his fellow-visionary blabs, A.E. notes with satisfaction that the recipient of the news was "astonished how I could keep secret such a thing" <**70**>. If actually said, this was perhaps intended as a criticism, but A.E. takes it as a compliment. On another occasion, when A.E. sees the ghost of the man who hid money, he tells another man who sees it "to say nothing" <**43**> - although, admittedly, in this instance the silence may have been to spare the feelings of the widow, and in fact a third party is soon told. The implication of secrecy on such occasions is probably that providence vouchsafes premonitions only to a favoured few. This was not necessarily merely A.E.'s view, for the first time he was involved in a visionary occurrence, a non-Catholic required him to keep it secret "untill that should be fulfiled which was showen" <**19**>. But A.E. seems to have held the view that providence especially entrusted glimpses beyond the veil to Catholics, hence secrecy involved keeping it in the family. The point is most clearly evident when the first Heaton vision is only vouchsafed to A.E. and his fellow-Catholic, John Stuart (who can be traced in the Catholic registers [CRS 1936, pp.197,303,310]), while two companions who are not stated to be Catholics but who argue over Scripture, and therefore are protestants, do not see the vision <**70**>. The second Heaton vision was also seen by A.E. and his fellow-Catholic <**71**>. A third instance of Catholic superior sensitivity to supernatural messages was when the ghost of the man who hid money was also seen by A.E.'s companion, James Trumble. For although A.E. does not otherwise identify Trumble, a man with that name was at an earlier date listed in the Newcastle Catholic register [CRS 1936, p.229]. Less tellingly, the nightwatchman's spell was observed by A.E. and his (Catholic) brother (but the name of the nightwatchman has not been found in the Catholic register - and it is uncertain whether the fact that it was Whit Sunday has any special significance) <**30**>.

Undoubtedly A.E. claimed certain occult powers for himself. "This Power of seeing a Spiret, Man or Woman, that is born in the twie light before the sunrise in the Morning, this was the case with me and Ann also" <**31**>. The first Town Moor vision was seen towards midnight (by the light of the moon?), and A.E. carefully specifies that the second was seen after sunset <**70-71**>. Other magic also worked only after dark. The nightwatchman's anecdote is headed "Before Sunrise in the Morning", and if the title is rightly placed it seems to imply that the spell worked only out of daylight hours. <30>. When A.E. was twice chased by fireballs, it was, first, after dark, and secondly, towards midnight <44>. But the more specific occult law, that ghosts of those born "in the twie light before the Sunrise" could be seen only by the favoured, was credited by A.E. as having been enunciated by a powerful authority - "This from the Reverend William Warlow, Cathlick Preast, Newcastle on Tyne" <31>. Whether in fact this is precisely what the ex-Jesuit Father Warrilow said to A.E. is perhaps capable of being doubted, but that the priest had a conversation with A.E. on the subject of ghosts which did not impress on the layman any scepticism about their existence is distinctly credible. It is true that Father Warrilow was, by later notions, a figure of his time. A newspaper item mentioned him in the following pithy anecdote. "As the Rev. Mr Warrilow, Roman Catholic priest, of Newcastle, was going into the boxes of the theatre of that place, a man attempted to take the watch out of his pocket ... Mr Warrilow, having procured a constable, he was apprehended and ... tried ... and sentenced to be transported to Botany Bay, where he rose to considerable eminence" [Sykes 1833, 1:347-8]. But if Father Warrilow was a man of his time, then his views regarding ghosts were probably not uncommon among his fellow Catholic priests.

Finally, as pointed out in the General Introduction, A.E.'s butter-setting magic consisted of words and a Catholic gesture, making "the sighn of the Cross". The words were "Depart from Mee, all ye that work Iniquety" and appear to be directed at the Devil (or his agent, a witch?), since A.E.'s mother subsequently thanks God that "thee hast Menifest thy Power above the Enemy" <21>. (Later in life, A.E. prays to be delivered from the Enemy, on the occasion when an atheist challenges the Devil and hence is given a rough night - "I hardly have a whole bone in my body, I have been over tree tops and thorn hedges, and all the Close that I had on is torn to raggs" <44>.) To make the butter set, A.E. exorcises the Devil; and exorcising was a service to the community practised within the Catholic community, normally of course only by priests, who did this fairly regularly [Bossy 1975, p.266]. While too much weight should not be laid on this last argument, since it appears that when a witch put a curse on butter-making even protestant cunning men could sometimes cope with the dairy magic, nevertheless it is difficult to avoid the conclusion that A.E. gained a reputation for magical butter-making because he was a papist - and he clearly believed that this was so. It is less clear that his other beliefs about the power of occult forces bore a close relationship to his Catholic beliefs and practices, but a case can be made out for thinking that this may well also have been so.

"The reason of my writing" ?

Even from the autobiography - on the assumption that A.E. repeats conversations correctly - it is evident that non-Catholics of the period were wont to invoke providence, or at least in have it on their lips in their more superficial discourse <**48,51**>. The General Introduction has argued that the most militant non-Catholics A.E. is likely to have encountered in his daily life were the Methodists, and that claims of providential dispensation were particularly strongly (and sincerely) expressed by the Methodists of the time. By 1805 there was a Methodist chapel at Felling Shore and some time later another at High Felling [Mackenzie 1834, 1:24]. A class-leader of the Methodists at Felling was killed in the 1812 disaster, and a pamphlet, based on a religious journal he kept, was instantly published [Lessey 1812]. The journal contains this typical entry. "I have been for three days at work, and I was unexpectedly discharged, for what reason I know not; but this I know, everything will work together for good to them that love God". Thus the Felling Methodist expressed the same sentiment of confidence in divine providence as the Catholic A.E. was to do in the opening paragraph of his autobiography. Fawdon Colliery, where A.E. worked during the 1810s, was in that decade notably a Methodist stronghold [Welford 1879, p.99]. Although A.E. never mentions the Methodists, or gives the explanation himself, the General Introduction has put forward the hypothesis that it was A.E.'s desire to out-Methodist the Methodists that generated his autobiography. Thus he wished to prove that the divine favour was not limited to protestants, but could extend to the papist minority, and that in fact his own life demonstrated that some Catholics might be especially favoured. The period of composition of the autobiography exactly coincided with a novel outburst of public assertiveness on the part of the local Catholic community. In 1823 a

Catholic Friendly Society was established at Newcastle, and in 1825 a Catholic Defence Society [Mackenzie 1827, pp.564,574]. If all this is correct, then however little A.E. says about his religious affiliation in the autobiography, his Roman Catholicism was fundamental to its production.

HEWORTH OLD CHAPEL.

COMMENTARIES

THE EDUCATION AND TRAINING OF A.E. AND HIS FAMILY

One of A.E.'s proud claims on behalf of his father was that "he gave us all an Education so as to make us fit for buisness" <2>. The only schooling we are told about is A.E.'s own, and we may doubt whether, in this narrower meaning of 'education', "us all" included A.E.'s sisters. (But see the Appendix to the Commentary on Family, where the possibility is raised that one of A.E.'s sisters came to own a girls' school.) Awareness of the social aims of pre-school education was shown by the Erringtons - the infant A.E. trailed his toy waggon, the adult A.E. commented that his father was "Ingrafting in me" his own trade of waggonwright <**4**>. At an unstated age, A.E. was sent to a local dame school where he "closely attended to read" - but not to spell with any accuracy, as confirmed by his own anecdote about how he excelled in class by spelling out "strainger", sic <**4**>. Despite being given the strap for mispronunciation - the poor child had a lisp - and later the cat-of-nine-tails at least twice, once in error, A.E. bore no grudge against his school-mistress and spoke highly of the dame, who was "loved by all" <**4,9**>. He gives no indication of how much time per week was spent at this school, or how long he continued at it. But "shortly after" learning to read he left the school, and some time after that the school-mistress died - this was in 1787 and A.E. was then aged eight and a half <**9**>. After dame school he went to what was most likely a more formal institution, "to Wm Yollowly of Low Huerth to learn figures in arithmetic ", probably in fact a parish or church school, since he refers to the involvement of the incumbent of the church <**5**>. Ascension Day was a school holiday, but this was perhaps to allow the boys to see a local spectacle on that day rather than to allow them to attend church <**10**>. At both schools, all the pupils mentioned are boys. Several anecdotes refer to A.E.'s time at the Heworth school <**6-8**>. He may have attended there daily, and unless there was a gap between dame school and this school, he must have been there for a period of some years: he must have started when he was not more than eight and he was certainly there during his twelfth year <**13**>. He implies that the master was a responsible citizen but otherwise does not commend him <**7,8**>.

When he reached the age of 13, his father decided that he should leave school and learn a trade. But when a week's trial at being a blacksmith proved distasteful, his mother intervened, suggesting a disagreement with A.E.'s father and providing a glimpse of an obscure figure in a forceful maternal attitude - "she thout it did not agree with me and I returned 'to School" <13>. If the anecdotes are in chronological order, A.E. combined his last period at school with work at night, thus being a 'part-timer' <14>. He was finally apprenticed to full-time labour at the age of 14, and so had had a schooling which, whatever its intensity, had been spread over several years of his life, certainly not less than six. Since it is virtually certain that his parents had to pay for all the schooling their children received, and even if only his three older brothers received the same privilege as A.E., the Erringtons in the 1780s were already displaying that self-sacrificing enthusiasm for schooling which was to become increasingly widespread among urban manual workers during the first half of the nineteenth century - an enthusiasm which preceded state intervention and the making of 'education' compulsory, universal, and free.

An early experience of the dangers of pit-work having scared him, A.E. was allowed by his father to try out a second surface job, that of farm labourer <15>. Robert Errington had himself been a surface worker in the early days of his working life, "going to the woods to woork with the Axe and Saw"; A.E.'s grandfather had been a farmer; and Robert retired from colliery work to "a small farm" <2> - hence the Errington family history testifies to the recruitment of the Tyneside mining industry from the neighbouring agricultural community. If Robert's willingness to let his son try out rural occupations underlines the contemporary symbiosis between mining and farming, A.E.'s discovery that surface jobs have their own unpleasantnesses (not least the exposure to a cruel surface climate) helps to explain the ease of mining recruitment, despite the perils underground. A.E. disliked farming and chose to return to pit-work.

A.E. tells us little about the education of his own children. But it seems that his oldest son only started work - or at least full-time work - when he was aged 13. Because pit-work gave him a serious illness, A.E. then apprenticed him instead to a surface trade <56>. A.E.'s second son was "at School" at a date between 1811 and 1814, when he was probably aged 10-12 <60>. Whether his daughters were given schooling is doubtful, since his youngest daughter could not sign her name on her father's death certificate in 1848.

A.E. was formally bound apprentice when he was 14 <17>, and in middle life he boasted that he had "served 7 years a trade" <57>. But as he was apprenticed to his own father, it is difficult to know what exactly this amounted to, apart from the fact that he worked with his father and picked up the trade. He married when he was 20, that is, before a seven-years apprenticeship was concluded, but his father was elderly and retired the next year. (A.E. only specifies his retirement from underground working, but given his age, and since he retires to a farm and no more is said about work at the colliery, we may assume that he retired totally from colliery employment). The impression is conveyed that A.E. was working on his own and earning a separate full wage before the apprenticeship was completed. At the beginning of the autobiography A.E. states that one stage of his life has been "that of Jurniman" <1>, but he never subsequently mentions the term or identifies the period. When it came to the turn of A.E.'s eldest son to begin work with his father, we are not told about any apprenticeship, perhaps because he was only 13. Later, he was apprenticed "to be a Painter, plumer and guilder for 7 years" at Newcastle <56>. Thus, like his father, A.E. did not force his son to continue in the trade he first chose for him. But whereas A.E. had tried out rural trades, the surface trade young Robert joined was urban, not rural - and partly indoors.

A.E.'s brother, John, also worked in Felling Colliery <17,30>, most probably as a waggonway wright, and if so, no doubt preceded A.E. in learning the trade from their father. But A.E. tells us nothing about this. Nor does he detail the training of his second son, Anthony Fenwick, who in 1818 was working with A.E. in making waggons <77>, and who may have been working as a waggonway wright with his father, perhaps part-time, for some years - his age in 1818 being probably between 16 and 18.

A.E.'S CONDITIONS OF WORK AND EARNINGS

Conditions of work

Whereas the process of coalworking required most of the underground labour force to be employed in regular shifts for a set number of days per week, it required the waggonway wright to be more flexible in his underground working hours, because his work involved frequent unpredictable and urgent circumstances as well as a regular routine. The shifts he regularly worked tended to be those within the maintenance period of the colliery week, that is, those during the hours of the night and during the day or days at the weekend when the pit was not drawing coal and was empty of most of its labour force. However, the waggonway wright had to fit in to his working week any time spent working at the surface - where presumably he did all of his waggon making and much of his waggon repairing, and where he might have to attend to surface waggonways - and also any additional hours spent underground when summoned there, or retained there, in emergencies.

A.E. notes some of this irregular working. His father was "3 or 4 nights a week ... down the pit" <**2**>, and as a young man A.E. was "frequently 2 or 3 nights in the week" underground <**14**>. In his early married life, A.E. came home from work at 2.00 a.m. <**31**>, and on a earlier occasion he went to work at 4.00 a.m. <**29**>. Work on Sunday nights was several times mentioned <**2,34,56,57**>. (This included Whit Sunday night, while on another occasion A.E. and his brother appear to be going to work in the early hours of Whit Sunday <**30**>, and he certainly also worked on Good Friday <**43**>. At this date church festivals were in general not recognised as occasions for industrial stoppage, and A.E. probably dates by them only because of his Catholic background.) Once A.E. mentions that on a particular Sunday night he and another man were the only workers down the pit <**34**>. A maintenance task, shifting an underground crane, was being done on a Saturday afternoon, at 2.00 p.m., when strange noises were heard, the more strange because the pit was not working coal and was therefore largely silent <**62**>.

The number of hours spent underground each shift was regulated for most workers, explicitly or implicitly, but for a waggonway wright was again more flexible. It is likely that he could, up to a point, decide the exact length of each shift for himself, the length depending on the work in hand and the other circumstances of the pit. Be that as it may, A.E. records that on occasions he worked shifts of many hours. One Sunday night shift began at 6.00 p.m. and lasted over 12 hours, A.E. returning home at 10.00 a.m. on Monday <**56**>; and twenty years earlier, when A.E. worked with his father on a Sunday night shift, they also returned home at 10.00 a.m. <**2**>. A twelve-hour maintenance shift was probably not uncommon, although the coalface workers generally worked a much shorter shift, nearer eight hours. A.E. records working 12 hours on 24 July 1812, a Friday, apparently at an emergency task <**56**>. But how many days in the week such long shifts were worked is not indicated, and probably long shifts alternated with much shorter ones. When a young man, A.E. is found returning to the surface at 3.00 p.m., together with a group of boys <**28**>. The presence of the boys suggests that this was the end of a coalworking shift, which probably began at 6.00 a.m. (Because underground coal conveyance took longer than coal-getting at the face, boys often worked longer hours than men). Even earlier, A.E. and his father left home for work at 6.00 a.m. <**20**>, but they were then doing work on the surface and therefore working daylight hours. However, on three later occasions A.E. records his time for going to work as 6.00 a.m. <**44,56,66**>. These shifts beginning at 6.00 a.m. must sometimes have coincided with coalworking shifts and presumably involved A.E. in inspection of the waggonways as they were running, but they were probably shorter than the maintenance shifts. It must be emphasised that A.E. merely refers to his working hours as matters of fact and at no point gives any indication of complaint about them. No doubt he accepted them, with other specific if irksome conditions of work, as part of the natural order, in those social aspects that appertained to men of his trade.

Only one episode supplies any evidence about a sequence of shifts. At Percy Main, one weekend A.E. works overnight Friday-Saturday and again Sunday-Monday, yet returns to work at 6.00 p.m. on Monday <**56**>. But this seems to have involved an emergency circumstance and was therefore probably not typical of A.E.'s normal working timetable.

On the Saturday in this sequence, A.E. slept after returning from work, he says for only for three hours, while on the Monday he rushed to Felling to see his sick son. But the three hours sleep was probably during the daylight hours of Saturday, and if so, he presumably also slept Saturday night to Sunday morning, something he omits to mention, in which case the impression given by the text, that over a period of four and half days he worked three shifts with only three hours of sleep in between is misleading. However, because of his visit to Felling, he apparently did miss sleep between two shifts. As has always been the case with coalmining, the fact that the work takes place out of the daylight, and hence can be done just as well at night, leads to the miners often working - and hence sleeping - 'unsocial hours'.

Early rising was common for workers of the period and their families. One Sunday morning when A.E. did not have to go to work, he was unable to sleep, so rose at 4.00 a.m., and it being summer he walked some miles to Heaton colliery <**70**>. His mother and sister began churning butter one morning at 4.00 a.m., and the housekeeper of an establishment nearby did the same <**20,22**>. When A.E. returned home at 2.00 a.m., his wife opened the door for him <**31**>. But they were newly-married, and so probably had no child yet to wear out its mother during the day, which may explain his wife's willingness to be disturbed in the middle of the night (even so, A.E. seems to have got his own supper). It is worth adding that A.E. remarks that "the fire was dull". 'Unsocial hours' were somewhat less painful for coalminers than they were for other workers because the miners' supply of free coal enabled them to keep fires on overnight.

Daytime Saturday, and to a lesser extent daytime Sunday, were times normally free from work, although some sleep might have to be fitted in to Saturday and work often began on Sunday evening. When A.E.'s surface adventures are ascribed to a particular day of the week, that day is usually Saturday, and it is likely that a number of the non-ascribed episodes also occurred then. On a Saturday he goes marketing to Newcastle <**41,45,47**> and visits his brother-in-law there <**52**>. He returns to Felling to see his children on a Saturday <**47**>, and almost certainly did this regularly. Late on a Saturday night he returns to Felling from Gateshead, where presumably he has been drinking <**54**>. Once his father went to Newcastle fair on a Saturday <**27**>; and once on a Sunday evening A.E. attended a party <**66**>. Nevertheless A.E. sometimes works on a Saturday or Sunday. As a boy, on a Saturday afternoon he helped his father with a surface job that needed to be completed <**18**>; and many years later, at Percy Main, he is found working underground on a Saturday afternoon <**62**>. At Gosforth, he went for a walk on a Sunday morning, but found colleagues doing an overtime job on the surface, and he makes it clear that he would have volunteered to participate in this Sunday work if he had been given the opportunity <**76**>. (As I understand the episode, A.E. had been told that the work would be done on a Sunday, but on a forthcoming one when he would be working, hence "it would go to our pay Setorday". But being done sooner, it was instead done by his rival, and hence went to "Marshels pay Setorday".)

How many days or shifts per week A.E. was expected to work is uncertain, nor is it clear to what extent his attendance at work was checked and regulated. But a Percy Main overman excused him a Sunday night shift <**66**>, and at Backworth, when he is invited to dine with his brother, William, he arranges for another man to do his work that day <**81**>. On the other hand, when he rushes from Percy Main to Felling to see his son, he also rushes back to begin work at 6.00 p.m. <**56**>.

"I got my money the pay night" <**79**>. A.E. refers to "pay Friday" at Felling, and to "pay Setarday" at Percy Main and Gosforth/Coxlodge <**65,76**>. An episode at Gosforth/Coxlodge has A.E. referring to his pay Saturday and another man's pay Saturday, indicating that fortnightly pay was paid out each week, but to one section of the workforce one week and to the remainder the next week <**76**>. On pay days A.E. probably collected his money before or after a shift, probably on a pay Friday before an evening shift, on a pay Saturday after an overnight shift. A.E. gives more detailed information about the payment procedure at Felling. "I was paid Evry 14 days. I had to go to the Ofice for the pay note, then I went to the Sighn of the Shakespear to get Change. I had 3 mens Money and my own" <**37**>. Payment to groups of workers rather than to individuals was not uncommon in this period, partly because employers found it difficult to get change, but sometimes because groups worked as contract gangs.

In this case it is not clear whether the group payment was only for reasons of convenience, or whether it means that A.E. collected a sum of money that covered the wages of his assistants as well as his own wage. Fortnightly pays were common, and A.E. states the amount paid to him at Percy Main "for 14 days" <**55**>, implying that there too he was paid fortnightly, as probably he also was at his later collieries. When he left Fawdon, he gave "11 days notice to quit" <**34**>, which may confirm a fortnightly pay (and presumably indicates a working fortnight of six days one week and five the next).

Both Erringtons worked long periods at Felling, as if on permanent contract, but they also occasionally undertook outside waggonway work, surely by agreement with their Felling employer when his work was slack and perhaps even at his request to oblige an acquaintance. A.E.'s father laid a waggonway in a ropery c.1799 and made waggons for another colliery (Old Fold) at an unstated date; while A.E. "made a railway at Sudick for Mr Wake" in 1809 (**2,45**).

Earnings

As a child, A.E. was paid 8d. per day for cleaning the way <17>, and thirty years later when his son started pitwork, he was paid 1s. per day for nine days' work, assisting the banksman at the surface <55>. As an adult, and just before he left Felling in 1811, A.E. earned a wage of 18s. per week - although he thought he deserved 24s. <55>. He moved to Percy Main, where in the first fortnight he earned £3.1s., or 30s.6d. per week <**55**>, so vast an increase that special circumstances must be suspected. His regular wage was probably nearer the Felling wage. He moved to Fawdon and in 1816 was paid 19s. per week, and he left Fawdon when this was reduced to 18s. <**68**>. The earnings of Tyneside colliers fell after 1815, broadly from £1 to 15s. per week, and the fact that A.E. seems to have more or less kept the same wage throughout the 1810s suggests that, as a skilled tradesman yearly gaining seniority in experience, he was doing better than most underground workers.

However, there is a complication to the assessment of his earnings position. A waggonway wright was either not always, or at least not solely, dependent on a fixed wage, because he might enter into a contract with the colliery. Normally the contract was to the effect that the waggonway wright paid out of his own pocket all of, or part of, the cost of constructing and maintaining a particular length of waggonway, and in return received a regular bonus whose size was related to the amount of use of that waggonway. A.E. certainly had a contract at Percy Main, as he explains. "The way was to be let afresh and I was cut out by One of my Men" - who bribed the overman with a pig - "thus I was decieved in my agreement" <**67**>. A.E. had apparently anticipated continuation of his contract, but it is not clear whether his complaint is that his contract was terminated unexpectedly and unfairly, or simply that it was not renewed as he had been led to expect. In any case, he sustained a heavy loss, because the maintenance agreement had not run long enough for the bonus payments to recoup his capital outlay. "I had laid 3 Inclines and found Nails and Candles for Score price which at the begining did not pay wages and I had to sufer loss. When I left I was £15 0s in det for nails and Candles and the totel loss was £30 0s." (It is worth noting that a working man could obtain credit to run up a debt of £30 - it is unlikely that A.E. can have found the sum out of personal savings.) The reference to wages may be to wages paid to the assistants that he had to find by the contract, or it may indicate that A.E.'s own wage had to come out of the maintenance bonus. The latter seems less likely, given his previous references to fixed wages. If indeed any contract payment was additional to a fixed wage, then it would be of interest to know how much A.E. earned this way, but he does not tell us; and if the sums were substantial, his total earnings may have been distinctly better than those of the average collier. But it is perhaps more likely that they were only marginally better. What he does tell us is that, working at Fawdon, he "in One year Redeamed £15 0s" of his debt, and did the same the next year <**68**>. Thus, in 1814-1816 his earnings were such that he could support his family and still save £15 each year, and this may indicate that his total earnings comprised more than a 19s. per week wage. He certainly had a contract at Fawdon, or at least was doing some piece work, since he tells us that in his early months there he earned £14 8s. by making, presumably on the surface and perhaps in overtime, "12 Chaldren Waggons at £1 4s per piece" <**68**>.

FOOD AND DRINK

Food

The main foods mentioned by A.E. are milk, cheese, bread and meat. His mother apparently kept cows since she made and sold "Old milk", butter and butter-milk <**20**>; and we hear of large houses nearby whose servants also made butter <**21,22**>. When a child, A.E. was given a piece of cake and a pot of buttermilk as a reward <21>. A poor girl living near Fawdon, on return home from a round delivering milk, was given as a meal only a penny roll and milk - so she said <**69**>. When A.E.'s household consisted of himself and his daughter they "got one pint New Milk in the Evening" <**69**>, but perhaps they also bought milk in the morning. Later, when temporarily teetotal, A.E., offered brandy after a dowsing, instead drank a pint of new milk <**82**>.

Cheese and bread, with beer or ale, was a common meal <**22,23,45,49**>. But when the Errington family offered cheese and bread to passing soldiers, they rose to "a cold table" for the officers <**23**>. Meat and bread was another stated meal. At an eating house, A.E. and his wife ate pieces of "warm meat" and bread <**45**>: elsewhere we hear of beef and biscuits <**16**>, a couple of biscuits and a slice of corned beef as a meal at work <**81**>, "beef stakes" and bread as a breakfast at a public house after night work <**64**>, and a slice of beef and mustard for a caller at a house <**69**>. There is no mention of potatoes or other vegetables, or of fruit, or of fish. We are not told how the cheese was obtained and whether the bread was home-baked. But we several times hear about the purchase of meat, because this was bought in large quantities at a central market. A.E. walked in to Newcastle to buy meat. Once three stone of beef was bought in the Flesh Market; another time A.E. and his neighbour bought beef, apparently in the evening, at 2d. cheaper than the morning price (and it was delivered by cart later - delivery of market purchases occurred again <**70**>). And once A.E. and his wife bought beef and mutton, including a steak, in the Butcher Market (the Flesh Market having been so renamed by the corporation - crassly, according to one pedantic contemporary, since the market sold flesh, not butchers [Mackenzie 1827, p.175]) <**38**,41,45>. We are not told how the meat was cooked at home. On the street penny pies were sold by a pieman <42>.

Meals at home included breakfast <**20,56,73**>, dinner in the middle of the day <**26,36,71,81**>, a tea of bread and butter <**69**>, and supper, a late meal, indeed once in the middle of the night, when A.E. returned from work - a dog also got supper <**30,31,59,73**>. The episode of A.E.'s double introduces us to two families at dinner on a Sunday c.1790, but all we are told is that "diner was on the plate" when one boy arrived and that the other boy's parents "dropd Knife and Fork" <**26**>. Meals at work were not described in any detail, apart from A.E. once eating "a couple of Biskets and a slice of Corned Beef" and another time sharing his supper with a young man <**81,66**>. (Sandwiches for the pitman's 'bait can' had probably not yet come in.) When a child A.E. went home to breakfast but took his father's meal to the pithead; and he also took his father's dinner to him when he was working on the surface railway <**20,11**>. Grateful employers twice promised A.E. and fellow-workers their supper, but how they provided it is not stated <**18,59**>.

Drink

Alcoholic drink is referred to over thirty times and clearly was a feature of most aspects of social intercourse, while "public houses" for the sale and consumption of alcohol are a common scene in the autobiography. Perhaps significantly, there is no mention of alcoholic drink being consumed underground, but exhausted rescue workers returning to the surface call for spirits <**70**>. As a child and youth, A.E. was given beer or ale as a reward, or as a medicine when sick - thrice specifically in public houses <**8,12,13,22,15,27**>. Gangs of workers completing lengthy tasks were rewarded as a group with "21 barrels of 8d bear", or with half a barrel of ale, to wash down cheese and bread, or with "strong bear" <18,49,74>. At public houses, a woman called for a gill of ale <**38**>: A.E. and other men usually ordered a pint of ale or beer <11,41,48,53,54,71>. But a gallon of ale was once donated to the house <**29**>, and on another occasion donated twice <**45**>; A.E. and a fellow tradesman drinking at mid-day had two quarts of ale <**47**>; and when A.E. met the pieman again, "the Brewer, hearing what pased, brought a full quort on the table to ingoy our selves, which we did" <**42**>. Elsewhere A.E. speaks more vaguely of taking "a glas or two" and "a glas a peace" <**41,51**>, or says that he "had a drink of ale and pased away the Evening" <**70**>. Hot ale and brandy was sometimes drunk, perhaps in winter: A.E. was rewarded with a quart by a grateful publican, and given one glass when unwell and dizzy, and then a second, as a medicine <**37,79**>. Invited into a house, he was given a glass of rum, with beef and mustard <**69**>. (He notes an exotic medicinal use of rum. Brother George, when a soldier in the West Indies, fought off fever by taking three pints daily for a week, then six pints daily for an unspecified period <**23**> - at least so A.E. says, and perhaps so George boasted.) For a good deed, A.E. was rewarded with two glasses of brandy (after earlier drinking a quart of ale) <**47**>; another time he was offered a glass of brandy <**45**>; and after getting very wet in the mine, once back on the surface the workers were told by their employer, who presumably paid, to "get brandy and injoy our selves" <**80**>. A party of women surprised A.E. by pouring brandy for themselves out of a tea-pot <**65**>. Applied externally, brandy also had medicinal uses <**28,56**>. Finally, when a celebration was held c.1820 after the exploration of the passages of Tynemouth Castle, "it was bear, and brandy, and rum" for the men and "jin for the ladies" <**80**>. There is no mention of other alcoholic drinks, unless when "Scotsmen treat[ed] us with Sperets" <52> the drink consumed by A.E. on that occasion can be assumed to have been whisky.

Drink sometimes accompanied meals. In A.E.'s parents' home, "we had diner and Mother set the botel on the table, and we ingoyed the afternoon in sochiel fellowship" <**26**>. Admittedly this was on a Sunday, and a visitor was present, but he was only a child. However, when A.E. and his wife had a steak as a day-time meal at a Newcastle eating house where drink was later served to others, it is not specified that alcohol was taken with the food; and the same was the case when A.E. ate "tea" at home <**45,69**>. Rather surprisingly, A.E. records his drinking tea at a public house, and two women doing the same at another; however the public-house tea-pot produced brandy for other women <**65,66**>. Celebration meals were of course drinking occasions. An evening at the 'Dog and Duck' for a dozen or so "tradesmen" lasted until 4.00 a.m.: "each drank what he liked and the cost was 6s to each, super and drink" <**29**> - the intake was apparently not overwhelming since some men proceeded immediately to work. After the Tynemouth Castle adventure, the party went to the 'Tynemouth Inn' - "And it was who was to be first with half crouns on the table. Each and Evry one served as they pased". But a group went on to the 'Wearmouth Bridge' for a private supper <**80**>. A.E. records only one drinking party at a private house, this being to celebrate his supposed engagement - "botels were brought on to the table to help our self" <**66**>.

Away from the public house A.E. often received free drink as a reward for services rendered, and within the public house drink sometimes came free to individuals, either as a result of a bet - "glases were bet" - or because it was paid for by one man as a penalty for starting, or at least losing, a fight <38,45,72>. Since A.E. was often a night-worker, his times of visiting public houses were probably untypical, but a few long evenings of drinking at public houses are mentioned <**29,68**, perhaps **79**>. On one occasion the evening's conviviality included singing by individuals, and songs by Burns were specified <**29**>.

Public houses

Public houses were of course centres for other than tipsy conviviality. Apart from the supper parties noted above, workmen had a solid breakfast at a public house <64>. This was at six in the morning, and A.E. also records a Newcastle pub being open before seven - on a winter's morning <63>. In this period the public house also provided an economic service. Change was in short supply and workmen were paid in groups with sovereigns or notes, requiring them to visit the only local source of change, the public house, in order to receive individual wages. A.E. documents this service with a good story <**37**>. He also documents two occasions on which the clientele of public houses contributed generously to a collection for distressed beggars, as well as the occasion on which they clothed the pseudo-Burns <29,63,64>.

The public houses mentioned were the following: at Newcastle, the Three Bulls Head <**25,42**>, Red Lion <**26**>, Dog and Duck <**29**>, Bay and Barrel <35>, Pack Horse <51>, Cock <52>, Hare and Hounds <65>, the

Low Crane House on the Quay <**63**>, and "the Fawdon house" <**71**>; at Gateshead, the Black Bull, Blue Bell and Half Moon <**24,47**>; at Felling, the Shakespeare and Unicorn <**37,50**>; at Howdon, an unnamed public house <**63**>; at North Shields, the Wearmouth Bridge <**66,80**>; at Tynemouth, "the Tynemouth inn" <**80**>; and at "Bublok" the Three Horse Shoes <**13**>. "Mr Barker's closet" and the eating house in "Dowrie Lane" <**41,45**>, both at Newcastle, may also have been basically pubs, although in 1811 'eating houses' were specifically distinguished from 'public houses' [Directory 1811].

The majority of these establishments can be traced in other contemporary records. The Three Bulls Heads in Castle Garth, the Dog and Duck in the Flesh Market, the "Cock, head of the Side", the Hare and Hounds on the Quayside, the "Low-crane house" also on the Quayside, all at Newcastle, and the Half Moon on Bottle Bank and the Black Bull in Black Bull Yard, both in Gateshead, were listed in local directories for 1787 and 1811 (in either or both), each being the only listed public house with that name in the appropriate locality. But in other instances the directories list two or more public houses with the same name in the same locality. Of the four Blue Bells in Gateshead, the one with a dwelling house standing opposite to it which A.E. could see from the Tyne Bridge <**47**> may have been either the Blue Bell in Pipewellgate or the Blue Bell in Bottle Bank - local experts could perhaps decide which is the more likely. In Newcastle, A.E. was perhaps more likely to have visited the Pack Horse, foot of the Side, than the Pack Horse in Pilgrim Street, a middle-class locality. His Sunday stroll past the Infirmary to reach a house opposite the Red Lion <**26**> probably took him to Newgate Street rather than to St John's Street - unless, of course, he chose a deliberately long way round to the latter, from a starting point in the Close, rather than a shorter way via Westgate Street, on the grounds that the latter course would have taken him past the chapel he should have been attending.

Other public houses are not listed in the 1787, 1811 or 1833 directories; however, the Unicorn at Felling Shore is shown on the 1856 OS map. It is just possible that the "Bay and Barrel" in Newcastle was the same as the Bay Horse, of which there was one in Pilgrim Street and another at Barras Bridge; but there is no mention in the directories of the Wearmouth Bridge at (or near) North Shields. (The name 'Wearmouth Bridge' would be more appropriate on the Durham side of the river, at South Shields, and on each of the two occasions the name is mentioned it is conceivable that the party involved in the episode crossed the river from North Shields to reach this public house - but in fact the name 'Wearmouth Bridge' is not recorded for South Shields either.) The "Fawdon house" and the "Tynemouth inn" are too loosely named to be identified. The three eating houses listed in 1811 were in the Flesh Market, but "Mr Barkers closet" is puzzling because none of the tradesmen named Barkus (there was none named Barker) listed in any of the directories was connected with the food and drink trade.

Of these various establishments, only a few appear to be represented by modern pubs of the same name on apparently the same or an immediately nearby site [Yellow Pages and telephone directories 1987]. The 'Half Moon' in Gateshead, an old inn which gave its name to presentday Half Moon Lane [Manders 1973, p.339], still exists, as does the 'Shakespeare' in Split Crow Lane (former High Felling Lane), Felling (but in A.E.'s day located between Gateshead and Felling). But the Black Bull in Gateshead, an old coaching inn, was swept away in the mid nineteenth century [Manders 1973, pp.103,240]. Other public houses may have migrated a short distance, for instance, the Blue Bell in Gateshead from Bottle Bank to High Street and the Hare and Hounds on Newcastle Quay several hundred yards east to Raby Street. The other public houses appear to have disappeared from anywhere near where they were located in A.E.'s day, but no doubt local historians could track down their period of existence from newspapers and local government records. None of the Newcastle public houses mentioned by A.E. was among the "respectable inns" listed in 1827 [Mackenzie 1827, p.719]; although the Cock, which disappeared in the 1890s, was in fact an old coaching inn [Middlebrook 1950, p.261]. A.E. also mentions visiting at least once the Freemasons' lodge in Gateshead, which met in a public house <**44**>.

Temperance

An inevitable consequence of heavy drinking is inebriation. On pay day, one particular collier only "left off drinking at 4 Oclock and was meary", and had to be helped home **<79>**; other individuals, a man and his wife, and a man "subgect to get two much", all got "two mutch" and had to be rescued by A.E. from death in the snow **<43,54,73>**. The women drinking brandy from the tea-pot were all "in Capable of walking alone" and the leading lady had to be supported in and out of the boat taking her across the river and home **<65>** - A.E. sounds mildly disapproving, possibly because these drinkers were women. He never admits to having been drunk himself, although his behaviour on a few occasions, even as he tells it, raises suspicions; and when his daughter is unkindly told that she has a drunken father, "this praid upon my mind and I made my daughter a promise I would drink no more" - a vow glossed as "not to touch speret or bear or small bear for one year". He drinks milk for a while, and when he terminates the period of abstinence, excuses himself on the grounds that "as a medison I had taken it" <79>. This episode occurs in the late 1810s and conceivably indicates some influence of the early stages of the temperance movement. But A.E. makes no general indictment of the Demon Drink, not even when noting violent behaviour in public houses - a fisticuffs challenge <45>, and assaults, in each case by four youths, first on A.E. himself, and then again, more viciously and apparently only for kicks, on an old man <41,45>. (In fairness, it is not clear from A.E.'s account whether the miscreants were under the influence of drink or were just commonplace youthful hooligans.) When A.E. says that his father "loved to be in Sochiel Company, and all ways endevered to restore peace when any frocthan [= friction] took place" <2>, the reference is most probably to public house brawls.

HEALTH

A.E. died when aged 69, his mother when aged 60, his father when aged 86 <2>. But his first wife died when aged 41, the cause not stated <46>, a sister died "of the Warter in the Brain at 11 years of age" <2>, and a daughter died aged 14 - "at that time a Fever was raging and She took the Fever and died in 2 days" <68>. However, at least six of A.E.'s siblings reached, or appear to have reached, adulthood (and A.E.'s statement that his parents "Brought up" their children may mean that eight reached adulthood), at least four appear to have reached middle age, and one, the only one whose age at death is known, died aged 58 <23>. Of A.E.'s own nine recorded children, the death of only one is noted (as stated above, at age 14), two certainly reached adulthood, the evidence suggests that three more almost certainly did, and there is no evidence either way about the remainder. The record is inadequate for firm conclusions, but the impression gained is that A.E.'s family was healthy - and fortunate. This impression is originally conveyed by what A.E. says of his father - "he was very healthy to the age of 76" <2>. The Erringtons do not seem to have incurred that level of mortality which was typical in Britain when A.E.'s father was born, and which was still not uncommon in certain rural and industrial areas when A.E. died. Whether the mortality experience of the Erringtons was typical of contemporary Tyneside is an open question. Outside his own family, A.E. noted that a miner who was killed left "a wife and 7 Children, al girls" <34> (he was aged 40 [Heworth parish register]), and a young man he once met "had an Orfin girl 8 or 9 yeares of Age which he Suported" <64>.

A.E. records only a limited amount of non-fatal illness. As a child aged two, he was "efected with the agoo" [? ague], presumably a fever, and then was very ill after being "efected in the small pock and being blind 9 days"; while shortly afterwards he had whooping cough <3>. As a result of a schoolboy accident, he lost the sight of one eye for six weeks <7>; and when an adult, he had sore shoulders from catching a falling man <47>. Between the ages of 76 and 86 his father was "efected with Rumatis at times" <2>. His brother, when serving as a soldier in the West Indies, for six weeks had "the sweting sickness", a disorder from which hundreds of his companions died <**23**>. Other disorders in the family resulted from occupational hazards and are discussed separately. Outside the family, A.E. refers to a boy poisoned by a slow-worm <**9**>, to a child whose neck was dislocated by a fall <**36**>, and to two beggars, "a Lamed Man" and a lad "Lame in boath feet" <**64**>.

In pit-work, A.E.'s father had his collar-bone broken when waggons ran away underground, and ribs broken on six occasions in other accidents <**2**> - a high rate of minor disaster which did not prevent him from living to a great age. A.E. himself, when a young man, sustained a wound on the shoulder in a shaft accident <**28**>, and when middle-aged "imbibed gass" at Walker Colliery and was ill for three weeks <**74**> - the gas left him chesty, certainly for a short time, perhaps for the remaining decades of his life. If this was all he suffered, compared to his father he got off lightly. And of course other underground workers lost their lives - a man burned in a minor gas explosion in one Felling pit from which another burned man recovered <**34**>, three men drowned by an inrush of water at another Felling pit <**43**>, two men burned in a minor explosion at Backworth <**78**>, a viewer killed at Hebburn <**65**>, "Some Men burnt and One his Leg broke" in an explosion at Percy Main, after which "One boy fell down the pit in Coming up" and was no doubt killed <**61**>. Finally, there is a reference to the 1815 Heaton disaster which cost the lives of 75 men <**70,71**>. Not mentioned but inferred is the 1812 Felling disaster which killed 92 men and boys <19>. A.E.'s earlier working life, from the 1790s to the 1820s, coincided with a period when a substantial proportion of the underground labour force in Tyneside collieries lost their lives in accidental violence, the mortality probably approaching 10% each decade [Hair 1968; Flinn 1984, pp.413-9]. Apart from the non-fatal accidents listed above, A.E. mentioned other occupational damage to health - an overman "sore burned, face and hands and breast" <**14**>, another man burned on another occasion <**34**>, and A.E.'s own son being "14 weeks down of the fever", as a result of working in foul water underground <**56**>. Again, a man lost his thumb, but this was the result of a malicious prank <**39**>.

The recorded treatment for all kinds of disorders was provided by folk medicine, and no fully professional medical man is mentioned in the autobiography. As a child A.E. was cured of whooping cough when he had medicine from "a man riding upon a pye bald gallaway who Brought me 2 peney worth of Suger Candy and Ordered Cream and Cours Shuger 3 times a day which soon restored me to health" <3>. His school mistress was "a good Doctriss, scield in Leting Bleed and Driving out venum" <9>. Later in life A.E. encountered "Dr Anderson" of Gateshead, a "pill docter" who was "always travelling with pills ... an honest man and well respected in the Counties of Durham and Northumberland" - despite being "subgect to get two much" <40,73>. Other medicine was administered or recommended by relatives and friends - "the white of an egg and Brandy" when A.E. had small pox <3>, bathing a blind eye with cold well-water <7>, "a Cup full of Brandy and Loaf Suggar" for a wounded shoulder, as provided by a kindly woman bye-stander <28>, washing with brandy to halt a feverish condition <56>, and consuming hot ale and brandy when dizzy <74>. Quite drastic, and perhaps apocryphal, was brother George's self-administered treatment for sweating sickness, three pints of rum daily for a week, increased to six pints per day, "sweting day and night" (hardly surprisingly) <23>. A.E. himself claimed a reputation as a practitioner of folk-medicine. After dealing with the dislocated neck of a child, he "then got the name of Cleaver Anty" <36), that is, Ant[hon]y the Bone-setter.

CRIME

In the 1780s there were "several people Robed on the way from Newcastle to Shields and Sunderland, the roads being infested at night by foot pads" <**8**>. A point on the Sunderland-Gateshead road continued to be known as 'Robbers' Corner' <**54**>. In the 1810s, walking home to Howdon from North Shields, A.E. was stopped by two men who attempted to rob him of the money he carried, ls.6d. <**59**>. But it is only fair to note that although A.E. travelled regularly around Tyneside on foot, sometimes for longish distances on lonely roads, this is the only instance of his having been mugged that he reports. He does, however, record two instances of minor non-violent thefts that came to his notice - a woman caught stealing a basketful of coal from a colliery <**30**> and a man who was alleged to have stolen A.E.'s son's dog <**60**>. Further, he once suspects two strangers of being 'resurrection men', that is, of being engaged in stealing newly-buried corpses from a churchyard, to sell to anatomists <**52**>. A.E. also once met a confidence trickster, the man who claimed to be Robert Burns <**29**>. Apart from the muggers' threats, violent criminal acts were reported thrice. The malicious prank that cost a man a thumb was deliberately planned (although wrongly executed) <**39**>, as was one of the two instances of assault in a public house, where assault was threatened against A.E. himself <**41**>. In the other instance, the bullying of an old man perhaps represented merely spontaneous youthful hooliganism <**41**>. Brawls were probably not uncommon, and A.E. twice commends individuals for being good at restoring the peace when frictions arose <**2,41**>.

Despite this being the age before the regular professional policing whose advent tends to be much applauded by historians, A.E. shows a casual confidence in the less regular agents of law and order of his day which suggests that they were not as ineffective as generally supposed. His finding of what he supposes to be a pistol is reported to "the Constable of the parish", who comments on a recent criminal episode <**8**>. When the dog-stealer, a man called Gordon "of a fiting Black arrecter", was reluctant to return the animal and became truculent, from the public-house where he encountered him A.E. "Called on a Constable to take Gorden and we would go Direct to A

Magistrate ... The Constable, a taylor by trade, took him in the Kings name as preasner". The man instantly surrendered the animal. The non-professional constable had lost part of his working day and so A.E. had to pay him 1s.6d. - most of which appears to have been instantly redistributed in drinks <**60**>.

But when assaults occurred in a Newcastle eating house, on each occasion "2 Sergents" were called, and they took the four men who had committed the assault into custody, the men being "handed off to the Kitty [= gaol]". A.E. was already acquainted with one of the Serjeants and said he had "seen him restore pease when a froction [= friction] took plase more than Once", and on the first occasion it appears that the Serjeants returned to have a drink with A.E. They told him that they would state the case before the magistrate and "would not request me to atend the Court on the Monday". A.E. does not indicate that he knew the outcome, and a suspicion is raised that the case was not pursued, probably because the men had bribed the Serjeants. However, on the second occasion, although A.E. again did not attend the court to give evidence, it appears that justice did take its course. "The Sergents came to see me and there was 5s for me for the asalt, and the Magistrate Said it was a Manly action to Save the Old man. They was bound over to keep the peace for 12 month with the old man and had all Expence to pay" <**41**>. These "Sergents" were among the seven Serjeants at Mace who constituted the full-time police establishment of Newcastle (in the 1820s on a salary of £225 p.a. each). Although progressive contemporaries alleged that "little attention was paid to age, activity or intelligence, in the choice of serjeants, who were, in general, ill qualified to discharge the arduous duties of their office" [Mackenzie 1827, p.626], A.E. seems to have been reasonably well satisfied with the existing police arrangements.

Finally, it is difficult to know what significance to derive from A.E.'s anecdote concerning the penny-pie seller <**42**>. With great trust and generosity, A.E. lends the man 14s. (since twelve pies had been bought and the seller had six shillings in hand, 13s. would have been enough to change a pound note) and the man later lends him 2s.6d. A.E. repays his loan eight years later. "I saw him no more for 8 years because he was transported to Vandemons Land for 7 years"

<42>. A.E. does not tell us for what crime the man was transported nor does he comment on the punishment.

COMMENTARIES

ANTHONY ERRINGTON - THE MAN AND HIS WRITING

Man and writer

A.E. tells us only a little about his physical appearance. A casual reference in an anecdote near the end of his autobiography reveals that he had "read hair" - to an extent that when he was aged forty or so it still attracted the attention of strangers <**80**>. His feats of strength suggest that he was a large, well-bodied man, a view possibly confirmed by the observation that children "of Anty's get ... had Strong bone in them" <**66**>. His first child was begotten when he was aged 19, the last of his recorded nine children when he was 52, he died when he was 69, and he records no serious illness after childhood, apart from a bronchial condition in middle life caused by inhaling gas. He thus appears to have been a strong, healthy man. But as a child he lisped, owing to "a small stopage in my speach" <**4**>; and one wonders if this impediment of speech did not survive into adulthood and explain part of his anxiety to win plaudits in other aspects of self-presentation. Throughout life he was familiarly known as 'Anty' <**26,32,36,66**>.

His mental characteristics are none too easy to assess, partly because they can only be judged from his own account of himself, told in select anecdotes. General aspects of his character have been discussed elsewhere (in the General Introduction and in the Commentary on Religion and Moral Values) - here discussion is limited to those characteristics which bear on the technical form of his autobiography. While his spelling is weak and he is so inexperienced in constructing narrative that he cannot translate his flow of thought into the conventional literary pattern of fully grammatical, sequential sentences, nevertheless he knows what he wants to tell the reader and he expresses himself at times vigorously and succinctly. He regards himself as having had schooling adequate for his station in life, and therefore for a trade which regularly required him to make simple measurements and calculations, and to write working notes and perhaps short reports and instructions to others. As we shall see, he seems to have kept some form of diary. He read the newspapers, both local and London papers <**35,41**>, and appears to have read at least one tract <**1**>. He received at least one letter, and wrote at least one <**35,56**>. His vocabulary is not wholly basic - he uses such words as 'motives','transactions','velocity','astronomy', 'pathetic','instrument','atheist','impostor','utility', 'perceived', 'interview','invincible' (although not all correctly). The reported speech in the account is often convincing, at least when it is simple and earthy - for instance, "Go Lad, and God be with thee" <**20**>, "Clag to, boys" <**28**>, "With your leave, Jentlemen, them that is needful cannot be mindful" <**38**>. But here and there rhetoric creeps in, particularly when A.E. wants to exalt the tone of an episode; and when odd rhetorical phrases resemble those in contemporary novelettes or melodrama it is difficult not to suspect the influence of some wider reading than newspapers - for instance, "in full Buty of Posesion of the River Tyne" <**10**>, "not one drie Cheek in the room" <**29**>, "not seen in the Anells of history" <**57**>.

Probably A.E. was no fool, in terms of sheer intelligence. Yet some of the anecdotes hint that he was a shade simple-minded emotionally, and was regarded as such by others. Perhaps he was seen as a mild eccentric, given to an undue measure of good works and discussion of the same. He shares in what appears to be the normal conviviality of his time and social grade, and seems to mix freely in company, and therefore can hardly have been an introvert, storing up his anecdotes for paper. But if instead he repeated them in company, he may well have set himself apart by this. Even the nickname he tells us he earned, "Cleaver Anty" <**36**>, while in one aspect complimentary, as indeed he saw it, in another aspect signalled that he was different. Whether his Roman Catholicism also set him apart is a very open question, since it is not clear how publicly he asserted it.

It is a fair point that for a marginally literate working man of this period to write his autobiography is a proof of some degree of eccentricity. Yet historians have recently learned that autobiographical writings by working men are less uncommon than was at one time supposed. A.E. was not the only Tyneside 'common man' to produce an autobiography in the period. The published autobiography of the York Minster incendiary, Jonathan Martin, has been noted in the General Introduction. Unpublished and lost is the autobiography of a man who, when a soldier in the Cheshire militia, and before he became a barber in Gateshead, achieved momentary headlines in 1796 by standing on his head on the top stone of the newly-built, 194-foot steeple of All Saints' church in Newcastle! [Mackenzie 1827, p.305]. In the 1830s a writer stated: "I have a memoir of him written by himself. In this piece of auto-biography he relates many hair-breadth escapes and various adventures which he has performed" [Sykes 1833, 1:381]. Thus, a barber of Gateshead was writing down autobiographical episodes at the same time as a waggonway wright of Felling.

A.E.'s autobiography consists mainly of a series of anecdotes, almost all of which refer to curious happenings. A.E.'s anecdotes are of course linked by a personal thread, and to some extent by an ideological thread, his 'hidden agenda', but in themselves they bear a close resemblance to a form of news item much favoured by the contemporary press and presumably much enjoyed by the contemporary reading public, the cautionary tale. For instance, the **Newcastle Chronicle** of the 1800s always concluded its column(s) of local news with several brief paragraphs relating dramatic, sensational and bizarre events, especially those events that (a) happened to otherwise unremarkable individuals, (b) terminated fatally or nearly so, and (c) enabled a social moral to be extracted and expounded. A selection of such news items formed a large part of a work first published in 1825, and significantly entitled **Local records, a historical register of remarkable events which have occurred in Northumberland and Durham** [Sykes 1833]. Such news items often began with the words 'A singular occurrence', and it is plausible that the term 'singular' which appears in the title of several of A.E.'s anecdotes was borrowed from the newspapers. It is unlikely that he saw the 1825 book but he certainly read the newspapers, and the inclusion in his autobiography of many anecdotes containing a bizarre element may represent his reflections after reading such news items, to the effect that strange and perilous things had happened to him too, and were worth reporting, together with, of course, their providential moral.

Dating the composition of the autobiography

When did A.E. compose his autobiography? He does not date most of his anecdotes. The only genealogical event dated is the death of his first wife, and here he gives the full date <**46**>. But he does not supply even the year-date of the births of his father's children, including his own birth, or of his own children, or the year-date of the deaths of his father and sister, or the year-date of his second marriage. Many contemporary protestant families would have recorded these events in a family Bible, with exact dates, but presumably the Erringtons did not. Yet it is by no means certain that A.E. is unaware of the dates he fails to supply, since he is capable of supplying dates for non-genealogical events, including one date in a period before he was born, the exact date of the commencement of Felling Colliery in 1777. He also gives the year-date of his father's working at a ropery and subsequent retirement in the same year, 1799 <**2**>.

In contrast to the paucity of dates for genealogical events, he supplies dates for a number of other events in his own life, particularly mining events. He supplies year-dates in the titles of two anecdotes, one a trifling episode in 1803 <**36**>, the other a more momentous event in 1813 <**57**> - and since he tells us that the latter episode occurred on Whit Sunday night, we can work out the exact date. He supplies the month and day of the 1815 Heaton disaster, but not the year, and the date he gives is wrong, strangely <**70**>. He manages to supply full dates for his engagement at Percy Main Colliery in 1811 and at Fawdon in 1814, and for his beginning work at Backworth in 1818, as well as for a celebration at Coxlodge in 1817 <**55,68,74,77**>. Other episodes of personal importance are also given full dates - the date in 1807 when the Discovery pit holed the waste, the date in 1809 when he saved a falling mason, the date in 1811 when he stopped a suicide attempt, and the date in 1812 when he and his son worked in water <**43,47,53,56**>. Extracts from what can only be working notes include three exact dates in 1800-1801 relating to specific work in progress down the pit <**32**>. It is clear, therefore, that A.E. kept a record of events in his life, at least from 1800. This record cannot have been confined to working notes since some of the events occurred on the surface, away from work. Presumably therefore he kept some sort of diary, perhaps however little more than a rough record of dates.

Many undated episodes can be approximately dated from information within them, especially from A.E.'s age at the time or his place of work. Occasionally, where only an approximate date can be deduced, an exact date can be obtained from the parish registers, for instance, the exact date of the deaths of A.E.'s sister and father. Parish register dates also enable us to fill out the family tree at points not mentioned in the account, for instance, the year of the death of A.E.'s mother. Doubts about the approximate dates of individual episodes exist in only two or three instances. Once the vast majority of the anecdotes have been given at least approximate dates, it becomes possible to examine the composition of the manuscript.

A detailed description of the manuscript is supplied in the final Commentary and it is sufficient to note here that the manuscript, as it is presently arranged, falls into two parts. The first part - fourteen sheets forming pages up to p.27 - appears as a continuous narrative, in the sense that 65 anecdotes continue down the whole of each page and then on the verso, and regularly run on from sheet to sheet, with only six starting at the beginning of a sheet. These anecdotes are broadly in chronological order (in the re-arrangement of the present edition only half a dozen anecdotes out of 65 have been judged to be out of chronological order, and the arrangement on the manuscript page generally shows that the fault was the author's, he having inserted them out of order). But the second part consists of eleven separate sheets, of different sizes and some only part-sheets; and normally each sheet contains only one or at most two of the total of 16 anecdotes, so that generally the anecdotes do not run on from sheet to sheet, and the verso pages are often blank. Although most of the anecdotes relate to the middle period of A.E.'s career (the final sheet with childhood anecdotes probably belongs to the earlier sequence), they appear in no specific chronological order.

Thus, whereas the first part was written as a narrative, mainly on sheets of the same size, the second part consists of sheets of different sizes apparently written in the main singly. The picture of composition that emerges seems to be this. A.E. composed the first part as a sequential narrative, covering the whole of his life up to the date he was writing. He then recollected further episodes within this period, which, as he thought of them, he wrote down on handy pieces of paper, sometimes only sections of sheets. He left the autobiography in this dispersed form, and it was gathered into a volume by a later hand, the loose single sheets being inserted after the sequential narrative, fairly haphazardly (and in one instance inserted into the narrative, quite haphazardly).

This implies that A.E.'s writing came to a halt in mid stream. Not only does the autobiography not cover the last two decades of his life, but it breaks off abruptly, since we may assume that if he had completed it up to the date of writing he would have signed off with a moral statement matching his opening passage. Moreover, in its course it announces two forthcoming episodes **<26,60>** which are not in the account as we have it. This could mean that some sheets have been lost from the narrative sequence, but it seems more likely that A.E. merely gave up. One possibility can be excluded. It cannot be that the whole account was first written down, at random, each anecdote on a separate sheet, and that then most of the sheets were sorted in more or less chronological order and copied to form the narrative sequence, leaving over a number of separate sheets; for the reason that the episodes on the separate sheets fall within the same chronological period as the episodes of the narrative sequence. There is, however, one curious distinction between the content of the narrative and the content of the remaining sheets. Whereas the narrative contains many underground episodes, the remaining sheets relate solely to surface episodes. This may mean that A.E. concentrated first on his mining career, perhaps because he had some written records, and then later recollected more surface episodes which he intended to incorporate.

Because we have no other record of A.E.'s handwriting we cannot decisively prove that the manuscript is in his own hand. However, any other possibility seems very unlikely. Since the narrative and the separate sheets are in the same hand, it cannot be that the narrative represents a copying up by another hand of anecdotes written down singly by A.E. Nor is it likely that A.E. merely dictated the anecdotes, since they incorporate working notes which must have been in his own hand - and there is firm evidence that he could read and write. It follows that the partly disorganised state of the autobiography, as we find it in the present manuscript, must be largely A.E.'s responsibility. He started the autobiography but left it uncompleted and partly in rough draft. This is a tidy conclusion, but problems remain.

When did A.E. compose his account? The episodes described in the autobiography can be dated with fair certainty, none occurring later than the mid 1820s. This correlates with the date at the head of the autobiography, 21 September 1823, which may well be the date when composition was begun. On the first sheet, A.E. refers to the death of his father, which occurred in 1818, indicating that the composition of the narrative as we have it (although not of any earlier drafts) began after this date. Hence it at first seems that composition proceeded from 1823 up to a date later in the 1820s. At one point A.E. dates an episode to 13 years after a meeting at Fawdon that occurred there before his daughter died: she lived at Fawdon and died there in 1814 <**69**>; therefore the episode relates to 1827. This episode, to be found in the narrative sequence, appears to be the episode occurring at the latest date.

But, in controversion of a tidy conclusion, casual references in other episodes - also within the narrative sequence - refer to later dates. The factory "at preasent" on the site of A.E.'s Low Felling cottage <**44**> was established in 1833. Brother George's death at South Shields <**23**> occurred in 1834. And finally, John Buddle, whose promise to employ A.E. "he fulfiled to his death" <**57**>, only died in 1843. Even if we ignore the last instance on the grounds that A.E. was being carried away by rhetoric and did not mean literally that Buddle was dead when he wrote, nevertheless it is certain that the other two references can only have been made in the mid 1830s, ten years after the autobiography breaks off. Two of the references appear at the very end of an anecdote and so might have been added at a date later than that of the general composition, but the third is not so placed; and examination of the manuscript text does not indicate that any of the three references has been added as a late insert. The conclusion seems irresistible, if somewhat puzzling. Despite the apparently strong evidence that the text was composed basically in the 1820s, the larger part of the manuscript we have -

which includes sheets watermarked 1812 - was written much later, certainly in the 1830s, perhaps even after 1843, in A.E.'s final years. (Whether this also applies to the separate sheets is uncertain.)

If the manuscript was written at a late date, could it be that composition was also much later than the 1820s - and perhaps that it broke off at A.E.'s death in 1849? It is true that at no point does A.E. actually state that he is writing in the 1820s - as he might have done, for instance, by referring to dated episodes as having occurred a given number of 'years ago'. It might also be advanced in favour of the late-composition theory that the narrative part of the manuscript has the appearance of untidy original composition, not the appearance of careful copying of a previous draft - but given the idiosyncrasies of A.E.'s style and writing this is not altogether convincing (nowhere does he correct errors or show indecision in expression). Alternatively, it can be argued that what we have, at least in the narrative, is a slightly revised copy of an earlier composition. That the reported episodes relate only to the 1820s and earlier is not of course proof that the autobiography was first composed in the 1820s. But the 1823 date at the head of the manuscript is difficult to interpret other than as the date when writing began - partly because it seems to refer to no specific episode in the account. Since it stands at the head of the narrative section, it does not rule out the possibility of earlier drafts, perhaps in the form of the separate sheets of the later section of the manuscript. But it makes it very unlikely that composition of the narrative was delayed beyond the 1820s, albeit the text was copied and slightly revised later, adding the post-1820s references. Other arguments for composition in the 1820s can be adduced. There is a flavour of relative immediacy and close recollection about the majority of episodes, those belonging to the 1800s and 1810s, which it is less easy to believe would have been present if the composition had occurred any later than the 1820s. There is also a 'dog in the night' argument. An odd feature of the autobiography is its failure to comment on steam locomotives and railways, although A.E. was virtually present at the birth of the system. This is very difficult to explain if composition occurred after A.E. had lived through the hectic and newsworthy development of the steam railway system in the

1830s and early 1840s, but much more explicable if he was composing in the early years, the 1820s.

The final conclusions must be to some extent tentative and even speculative, but it seems most likely that -

(a) A.E. began composition of the first draft of the narrative sequence in 1823 (although it is conceivable that he had earlier drafts of some of the episodes) and probably wrote it over a period of years, perhaps up to 1827, but without completing what he originally intended.

(b) He also composed the individual anecdotes of the separate sheets, either as he was proceeding with the narrative, in which case he failed to incorporate them in a first draft, for an unknown reason, or, more likely, after he had completed composition of the narrative, in which case he never got round to incorporating them.

(c) At a later date, perhaps in the mid 1830s, perhaps later, he copied the narrative sequence and slightly updated it, but still failed to incorporate the anecdotes on the separate sheets (perhaps he had mislaid them), and he also failed to add to the narrative (unless some part is lost), in order to bring the autobiography up to date.

COMMENTARIES

THE MANUSCRIPT AND ITS EDITING

The history of the manuscript

The manuscript of A.E.'s autobiography is in Gateshead Central Library, where it is listed as O/Errington 61/2. It came to the library from a local historian, John Oxberry, and was probably willed with other material at his death in 1940. All that we know about its earlier history is stated on a postcard which is inserted into the manuscript volume and which was sent to Oxberry in 1915 by another local historian, Richard Welford of Gosforth. Welford, an older man, had published on the history of Tyneside since the 1870s. The postcard states that he is sending the Errington manuscript to Oxberry, who may add it to his collection of local documentation, "if you think it worth keeping". As to its earlier history - "It turned up among some old papers yesterday. Where it came from I cannot remember." A.E. died in 1848, and it is possible that Welford acquired it only a generation later. Although in his old age he had forgotten the circumstances, it is likely that Welford had collected it because of its references to the Gosforth area - Gosforth parish formerly included Fawdon, Kenton and Coxlodge - on which he published a history [Welford 1879].

But neither in this work nor in his other writings does Welford appear to mention the autobiography; and Oxberry, whose writings included contributions on local history to local newspapers, equally does not appear to have made any use of the account passed on to him by Welford. In 1951, in the course of research into early nineteenth century coalmining, I came upon the manuscript in Gateshead Library and transcribed it. As far as I know, the autobiography was not listed or mentioned in print before note was taken of it in David Vincent's analysis of working class autobiographies [Vincent 1981].

Description of the manuscript

The manuscript in Gateshead Central Library listed as O/Errington 61/2 consists of 25 sheets of text, the largest sheets of extra foolscap size (380mm x 240mm). The sheets are paginated in pencil as

1-27,30-33,33bis-40, some verso pages being blank, and the very first page unnumbered; also, two sheets are unpaginated. These text sheets are preceded by two sheets that are blank apart from a label on one and an inserted postcard showing on both sides of the other. The text sheets have been pasted at one edge and secured to central strips, and the whole forms a thin stitched volume with plain, dark, limp covers, of nineteenth century appearance. (Most probably the sheets were stuck into an existing stitched volume of writing paper whose interior pages had been cut away nearly to the inside margin, leaving the strips.)

The table on the following page numbers the sheets, and relates them both to the page numbers of the manuscript and to the numbers and arrangement of the anecdotes in the present edition.

The handwriting of the manuscript is the same throughout, although there are changes of ink and differential fading of pages, the latter probably indicating that the loose sheets lay around for some considerable period of time in daylight, perhaps during composition. The full size sheets are watermarked, the others are not. Each watermarked sheet has one or the other of two linked devices, either a horn in a crowned shield (a common watermark device) with the letter W, or the letter W with a date, 1812 - undoubtedly the date of manufacture of the paper of all these full size sheets. Writing paper might be retained for decades, but equally the date 1812 is not inconsistent with composition of the manuscript, or at least of these sheets of it, in the 1820s.

The first twelve sheets, together with sheets 14-15 (even although these sheets are of a different size), carry what appears to be a single sequence of anecdotes, broadly in chronological order. Within the pages that are grouped in the table (i.e., 4-13,14-17,19-23, 24-27), the anecdotes run on from page to page. The final sheet of the manuscript is of the same size as the first twelve sheets and has probably fallen out of the sequence (it may or may not be significant that its anecdotes relate to A.E.'s childhood). The remaining ten sheets are not of the same size as the first sheets and are of a number of different sizes. These sheets appear to have been written separately and singly, that is, not as part of a narrative sequence.

TABLE : Make-up of the manuscript volume

<i> reverse of front cover (label, ex libris John Oxberry)
<**ii**> one sheet unn. page (label, 'Presented to Gateshead Public Library by John Oxberry F.S.A.') / blank
<**iii**> one sheet unn. pages, insert through sheet (postcard to Oxberry from Welford, 16.11.1915)

text sheets

	(page nos.)	(anecdote nos.)	(sheet size)
			(all full size)
<**1-7**>	one unn. + 1-13	= 1-9,11-14,10,15,17,16,18-19,24-28,32-33,31,34,23, 30	
<**8-9**>	14-17	= 43-44,46-47,49-50,48,51-54	
<**10-12**>	18-23	= 55-62,66-68,70,69,71,73	
			(various sizes)
<**13**>	two unn.	= 63-64	quarter length sheet inserted between <10> and <11>
<**14-15**>	24-27	= 74,76-78, 75,79-83	two sheets of slightly less than full length
<**16**>	two unn.	= 41	quarter length sheet
	[28-29 lacking]		
<**17**>	30/blank	= 41 cont.	one sheet of much less than full length
<**18**>	31/blank	= 35	one sheet of a length between that of <14-15> and that of <17>
<**19**>	32/blank	= 36-37	one sheet of another intermediate length
<**20**>	33-33bis	= 38,72,39	one sheet of the same length as <17>
<**21-23**>	34/blank	= 40	three sheets of quarter length
	35-36	= 42	
	37/blank	= 65	
<**24**>	38/blank	= 45	one sheet of the same length as <14-15>
<**25**>	39-40	= 20-22,29	one sheet of the full length

The postcard reads as follows. "Gosforth November 16,1915 / Dear Mr Oxberry, / I am sending you by book packet post a curious autobiography of a man who worked at Felling and other adjoining places on the Tyne at the beginning of the last century. It is easy to read when you once get into his curious way of spelling. Where it came from I cannot remember. It turned up among some some old papers yesterday. It is full of signs and portents and miraculous interposition. Pray add it to your collections if you think it worth keeping. / Yours very sincerely, / Rich. Welford."

Each sheet contains either a single anecdote or at most only two anecdotes. Generally the anecdotes do not run on from sheet to sheet or even page to page, and often the verso page is blank. (The particular anecdotes refer almost exclusively to the middle years of the autobiographical period, but the significance of this is unclear.) When the volume was assembled, these remaining sheets were inserted without regard for any recognisable chronological order and with little regard even for the size of the sheets. (Two sheets of the same size, 17 and 20, are separated by two sheets of different sizes, while the penultimate sheet, 24, is of the same size as the two final sheets of the narrative sequence, 14 and 15.) Thus, whereas sheets 1-12 are linked by size and with sheets 14-15 and perhaps 25 by sequential narrative, the other ten sheets were inserted more or less pell-mell.

The numbered pagination is in pencil and includes an error of omission and an error of duplication. It was probably added after the volume reached the library, where certainly the notice of its 'presentation' by Oxberry was attached. It is unfortunately uncertain when the volume arrived, but, despite the wording of the notice, it is likely to have been after Oxberry's death in 1940. The ex libris label on the inside of the front cover suggests that the volume reached the library in its present form, apart from the presentation notice and the numbering of pages. Again, apart from the ex libris label and the insertion of Welford's postcard, it seems likely that the volume reached Oxberry in 1915 in its present form; and since Welford's message does not indicate otherwise, it may well be that it also reached him in its present form. This was at an unknown date, which in view of his lack of recollection in 1915 was presumably before 1900.

If the above is correct, then the loose manuscript sheets were made up into a volume by an unknown hand in the nineteenth century. The lack of care in the ordering of the later sheets suggests that it was not A.E. who formed the volume, but someone into whose hands the sheets passed after his death in 1848, conceivably one of his children or grandchildren. The insertion of the quarter length sheets - which are probably sliced-off sections of a larger sheet - seems to have provided a special problem, and they were inserted mostly at random. The various disorderly features in the make-up of the manuscript as we now have it, some of them of uncertain significance, mean that the manuscript's present form throws only limited light on the way in which the autobiography was composed.

Editing procedures

The text of the autobiography is difficult to follow in its original form and has therefore been editorially modified in the ways described below. I hope that a sensible balance has been struck between total ease of reading and scholarship, inasmuch as some changes have been introduced silently while others are openly signalled. The signalling of changes is not mere pedantry. It is important that the reader should recognise that, here and there, the meaning of the original is obscure, partly because of lack of punctuation and A.E.'s habit of omitting words when a rush of thought overwhelms his speed of writing. All the minor changes I have made to the text are designed to make it more obviously meaningful, but certain forms of explanatory addition are signalled, because in some places I may have misinterpreted A.E.'s line of thought and the reader may be able to supply an alternative and better explanation. I would welcome any emendations of this kind.

Punctuation

The manuscript text regularly lacks any normal punctuation - even the employment of a capital initial letter on the first word of a sentence. Punctuation has been silently supplied, and therefore the text split up into commonplace sentences, although an attempt has been made to retain the colloquial, conversational flavour by having some sentences which are less than fully grammatical but whose meaning is crystal clear. Very occasionally, in a long breathless passage, the more radical step has been taken of silently reversing the order of two phrases or even sentences.

Verbal changes

(a) **silent changes** A.E. regularly omits the pronoun subject of a clause or sentence, and instead links a string of utterances with a series of 'and'. In editorially-delineated sentences the pronoun, which in fact is almost never in doubt, is silently supplied, while on the other hand many instances of a linking 'and' are silently omitted. But in very clipped passages, an 'and', or, in clause contrasts, a 'but' (a term almost never used by A.E.), and also 'then' and 'next', are at times silently inserted.

(b) **signalled changes** Any other verbal additions are signalled by being placed within square brackets, thus [ ]. Verbs have sometimes to be supplied, normally with fair certainty. But when an explanatory phrase has to be inserted to make sense of an obscure passage, doubt about the correctness of the explanation may occasionally arise, and the reader is invited to reflect. Another form of insertion relates to technical and dialect terms with which the general reader may not be familiar. These are glossed or explained in square brackets, the addition being introduced by an equals sign, thus [=].

Spelling

The hand of the manuscript is not difficult to read, only a very few words being fully or partly illegible, and some of these can be guessed with reasonable confidence. Where doubt still exists, this is signalled by a query mark within square brackets [?]. But the spelling is idiosyncratic, and words are not always consistently spelled or mis-spelled. Words mis-spelled in less than an obvious way are followed by the correct spelling in brackets. A few short words regularly lack a terminal letter and could therefore confuse the reader, notably 'the' for 'they', 'of' for 'off', and 'som' and 'com' for 'some' and 'come'. These words are silently extended to the correct spelling. Other mis-spelled words are presented as in the manuscript, but very occasionally, for clarity, a missing letter is inserted within square brackets.

Because of the frequent mis-spelling, it is sometimes difficult to be sure about individual letters, particularly vowels. In A.E.'s hand the letters 'a' and 'o' are especially difficult to distinguish within words, and unfortunately the pronunciation of these particular vowels is one of the features that sets the Tyneside dialect - which often guides A.E.'s spelling - apart from standard English pronunciation. Words containing these letters have been transcribed as read, ignoring the 'correct' spelling. But I must admit that my readings may sometimes exaggerate the idiosyncrasy of A.E.'s spelling, since in common words we all of us lapse into careless shaping of letters, not least vowels.

The manuscript text contains a multitude of words beginning with capital letters. A.E. was of course following a feature of contemporary spelling, but his use of capitals is excessive and shows neither rhyme nor reason. There is no grammatical consistency in the use of initial capitals, and they certainly do not indicate stress or any other semantic signal. Indeed, as initial letters of words the letters 'a','c','d','e','i', 'o' and 's' are so frequently written as capitals that it may be thought that A.E.'s hand regularly forgot the distinction between the capital and non-capital forms. If this was the case, then it could be argued that the text should have been presented without these meaningless capitals, which only serve to exaggerate the flavour of quaintness. On the other hand, A.E. often fails to supply initial capitals on names of persons and places, and also at the beginnings of sentences. To clarify meaning in such instances, capitals have been supplied in replacement, silently.

Order, numbering and titles of the anecdotes

As explained in detail previously, the first part of the manuscript presents a majority of the total anecdotes broadly in chronological order, but the remaining sheets contain anecdotes which fit within the previous chronology but are in no particular order. I have re-arranged all the anecdotes in chronological order, but sometimes only approximately in that order, either because some cannot be dated exactly or, occasionally, because I have retained A.E.'s order when, in the first part of the manuscript, he seems deliberately to have back-tracked on strict chronological order. The original order is given in the Table on p.232.

For easy reference in the scholarly apparatus, I have numbered the anecdotes. In both the text and the apparatus the numbers appear in bold print within oblique brackets, thus <**83**>.

A.E. supplies titles for the majority of his anecdotes. Where he fails to do this, a title has been silently supplied, the title being taken either from the first words of the anecdote (and not necessarily repeated) or from some phrase within the anecdote which serves to introduce it.

BIBLIOGRAPHY OF WORKS CITED
in the notes to the text and in the commentaries

Ashton and Sykes 1929 — T.S.Ashton and J.Sykes, **The coal industry of the eighteenth century**, Manchester, 1929

Aveling 1976 — J.C.Aveling, **The handle and the axe: the Catholic recusants in England from reformation to emancipation**, London, 1976

Bailey 1810 — John Bailey, **General view of the agriculture of the County of Durham**, London, 1810

Bailey and Culley 1797 — J.Bailey and G.Culley, **General view of the agriculture of the County of Northumberland**, Newcastle, 1797

Bossy 1967-8, 1969-70 — John Bossy, 'Four Catholic congregations in rural Northumberland 1750-1850', and 'More Northumbrian congregations', **Recusant History**, 9, 1967-8, pp.88-119; 10, 1969-70, pp.11-34

Bossy 1975 — John Bossy, **The English Catholic community 1570-1850**, London, 1975

Brand 1789 — John Brand, **The history and antiquities of the town and county of the town of Newcastle upon Tyne**, 2 vols., London, 1789

Campbell 1964 — W.A.Campbell, **The old Tyneside chemical trade**, Newcastle University, [1964]

Compleat Collier 1708 — **The Compleat Collier, or, the whole art of sinking getting, and working coal-mines, etc. as is now used in the Northern Parts, especially about Sunderland and Newcastle. By J.C.**, London, 1708

Craster 1907 — H.H.E.Craster, **A history of Northumberland, vol.8, The parish of Tynemouth**, Newcastle-upon-Tyne, 1907

CRS 1914 — C.J.S.Spedding, ed., 'Catholic registers of Capheaton, Northumberland', in **Miscellanea**, Catholic Record Society, vol.9, 1914, pp.237-248

CRS 1936 — J.Lenders and J.R. Baterden, eds. 'The Catholic registers of the Secular Mission in Newcastle-upon-Tyne which ultimately became St. Andrew's, from 1765', in **Miscellanea**, Catholic Record Society, vol.36, 1936, pp.198-324

Directory 1787 — **The Newcastle and Gateshead Directory for 1787,-89** [sic], [n.d.,n.p.]

Directory 1801 — **The Directory for the year 1801, of the town and county of Newcastle upon Tyne, Gateshead, and places adjacent**, Newcastle upon Tyne, 1801

Directory 1811 — **Mackenzie and Dent's triennial Directory for Newcastle upon Tyne, Gateshead and places adjacent**, Newcastle, 1811

Directory 1821 — **Second edition of the Commercial Directory of Ireland, Scotland and the four most Northern Counties of England for 1821-22 and 23**, Manchester, 1820 [sic]

Directory 1822 — **Pigot's Directory for Northumberland**, [Newcastle], 1822

Directory 1827 — W.Parson and W.White, **History, directory and gazeteer of the counties of Durham and Northumberland**, Newcastle, 1827

Directory 1833 — **A directory of the towns of Newcastle and Gateshead**, Newcastle, 1833

Directory 1834 — **Pigot and Co.'s National Commercial Directory ..., the counties of Chester, ... Durham, ... Northumberland**, London, 1834

Directory 1865 — **Ward's Directory comprehending the towns of Newcastle, Gateshead, Shields and their localities 1865-6**, Newcastle, 1865

DNB — **Dictionary of National Biography**

Egan 1818-1824 — Pierce Egan, **Boxiana, or sketches of modern pugilists**, 4 vols., London, 1818-1824

Flinn 1984 — Michael W. Flinn, **The history of the British coal industry, vol.2, 1700-1830**, Oxford, 1984

Fortescue 1900 — J.W. Fortescue, **History of the British Army**, vol.4/1, London, 1900

Galloway 1898 — Robert Galloway, **Annals of coal mining and the coal trade**, 2 vols., London, 1898,1904, repr. Newton Abbot, 1971 [only vol.1 cited]

Gibson 1849 — W.S.Gibson, **A guide to Tynemouth**, North Shields, 1849

Gill and Burke 1987 — M.Gill and M.Burke, 'Coal miner mobility: North East England in the early nineteenth century', **Journal of Regional and Local Studies**, 7, 1987, pp.35-54

GM — **The Gentleman's Magazine**

Graham 1986 — Frank Graham, **The Geordie netty**, Rothbury, 1986

Greenwell 1888 — G.C.Greenwell, **A glossary of terms used in the coal trade of Northumberland and Durham**, 3rd. ed., London, 1888 [first ed. 1849]

Hair 1844 — T.H.Hair, **Sketches of the coal mines in Northumberland and Durham**, London, 1839/1844, reprinted Newcastle and Newton Abbot, 1969

Hair 1965 — P.E.H.Hair, 'The binding of the pitmen of the North East 1800-1809', **Durham University Journal**, 58, 1965, pp.1-13

Hair 1966 — P.E.H.Hair, 'Bridal pregnancy in rural England in earlier centuries', **Population Studies**, 20, 1966, pp.233-243

Hair 1968 — P.E.H.Hair, 'Mortality from violence in British coalmines, 1800-1850', **Economic History Review**, 21, 1968, pp.545-561

Hodgson 1813 — John Hodgson, **The funeral service of the Felling Colliery sufferers; to which are prefixed, a description and plan of that colliery; an account of the late accident there; of the fund raised for the widows; and suggestions for founding a Colliers' Hospital**, Newcastle, 1813

Hughes 1952 — Edward Hughes, **North Country life in the eighteenth century: the North-East, 1700-1750**, London, 1952

Hutchinson 1787 — W.Hutchinson, **The history and antiquities of the County Palatine of Durham**, 3 vols., Newcastle, 1787

Kentish 1797,1817 — Edward Kentish, **An essay on burns, principally those which happen to workmen in mines**, London, 1797, 2nd.ed. 1817

Kinsley 1968 — J.Kinsley, **The poems and songs of Robert Burns**, Oxford, 1968

Kittredge 1928 — G.L.Kittredge, **Witchcraft in Old and New England**, Cambridge, Mass., 1928

Lee 1951 — Charles E.Lee, 'The wagonways of Tyneside', **Archaeologia Aeliana**, 4th ser., vol. 21, 1951, pp.135-202

Lessey 1812 — Theophilus Lessey, **A short account of the life and Christian experience of John Thompson, one of the persons killed by the explosion which took place in the Felling Colliery, on Monday May 25 1812. Compiled chiefly from his own journal**, Newcastle, 1812

Lewis 1975 — M.J.T.Lewis, **Early wooden railways**, London, 1975

Mackenzie 1825 — E[neas].Mackenzie, **A historical, topographical and descriptive view of the county of Northumberland**, 2nd. ed., 2 vols., Newcastle upon Tyne, 1825

Mackenzie 1827 — E.Mackenzie, **A descriptive and historical account of the town and county of Newcastle upon Tyne, including the borough of Gateshead**, 2 vols. in one, Newcastle upon Tyne, 1827

Mackenzie 1834 — E.Mackenzie and M.Ross, **An historical, topographical and descriptive view of the County Palatinate of Durham**, Newcastle upon Tyne, 1834

Manders 1973 — F.W.D.Manders, **A history of Gateshead**, Gateshead Corporation, 1973

Middlebrook 1950 — S. Middlebrook, **Newcastle upon Tyne: its growth and achievement**, Newcastle upon Tyne, 1950

Miscellanea 1930 — **Miscellanea**, Newcastle upon Tyne Record Committee, Newcastle, 1930

NEIME 1885 — **An account of the strata of Northumberland and Durham, as proved by borings and sinkings, part 2**, North of England Institute of Mining Engineers, Newcastle, 1885

RCCM 1842 — **Report of the Children's Employment Commission, First Report (Mines), Appendix, Part 1**, Parliamentary Papers 1842:XVI

Reid 1889 — J.B.Reid, **Concordance of the poems and songs of Robert Burns**, London, 1889

Richardson 1846 — M.A.Richardson, **The Borderer's Table Book**, Newcastle-upon-Tyne, 1846

Roy 1985 — G.Ross Roy, **The letters of Robert Burns**, London, 1985

Steel 1974 — D.J.Steel and E.R.Samuel, **Sources for Roman Catholic and Jewish genealogy and family history**, London, 1974

Sykes 1833 — John Sykes, **Local records, or historical register of remarkable events which have occurred in Northumberland and Durham**, 2nd. ed., Newcastle, 1833

Thomas 1971 — K.Thomas, **Religion and the decline of magic**, London, 1971

Tomlinson 1888 — W.W.Tomlinson, **Comprehensive guide to the County of Northumberland**, 10th ed., London, 1888, repr. Newcastle, [1923]

Vincent 1981 — David Vincent, **Bread, knowledge and freedom: a study of nineteenth-century working class autobiographies**, London, 1981

Vincent Smith 1930 — W.Vincent Smith, **Catholic Tyneside from the beginning of the Reformation to the restoration of the hierarchy 1534-1870**, Newcastle-on-Tyne, [1930]

Walsh and Forster 1969-70 — E.Walsh and Ann Forster, 'The recusancy of the Brandlings', **Recusant History**, 10, 1969-70, pp.35-64

Welford 1879 — Richard Welford, **A history of the parish of Gosforth**, Newcastle upon Tyne, [1879]

Localities of manuscript sources

DCRO = Durham County Record Office, Durham (parish registers)

GCL = Gateshead Central Library (Errington text, Taylor typescript, local maps and plans, local census material)

NCRO = Northumberland County Record Office, Gosforth, Newcastle (Buddle, Watson and Easton Papers, Gosforth parish register)

NCL = Newcastle Central Library (Northumberland and Durham parish register transcripts)

NSCL = Local Studies Centre, Old Central Library, North Shields (local census material, Taggart typescript c.1930)

PRO = Public Record Office, London (Roman Catholic registers, 1851 and 1861 census)

GLOSSARY

OF TECHNICAL AND DIALECT TERMS

Many of the terms listed below, being still in use, are glossed from personal knowledge, but are also to be found, together with a number of obsolete terms, in Greenwell 1888; or in Graham 1987, which is based partly on earlier dialect lists.

after damp — the asphyxiating gas (carbon dioxide) produced after a methane ('fire damp') explosion underground, also called 'choke damp'

amain, to run amain (said of waggons, etc) — to break loose on an incline and run down it out of control

arles — bounty or earnest money given to workman on accepting job

bairn — child

bank — the top of the mine-shaft, the surface

banksman — workman at the top of the shaft who regulates the movement of coal containers and men up and down the shaft

bargain men — workmen on competitive contract for supplementary underground tasks such as stone-working and rolleyway making

barrier — area of coal left intact between workings, especially the workings of different collieries, often to hold back water or gas

beat need, to be someone's beat need — to be his last resource, fall-back, stooge

besom — broom

blower — violent eruption of water or gas through a face or wall underground

bord (or board) — limited gallery or passage within the coal seam from which coal is being excavated, at faces in front or at the sides, and then passed outwards into headways

brake, to brake an engine — to operate a pumping or winding engine

GLOSSARY

brattice	partition of wood or cloth to direct or stop the flow of air
burn	brook
carrying of the air	see 'coursing'
chalder (or chaldron)	measure of coal, the Newcastle chalder being the equivalent of 53 cwt.
clag, to	adhere to
cleaver	bone-setter
convoy	hand-brake on coal waggons
corf	wickerwork basket used to carry coal from the face and up the shaft, later replaced by tubs
course the air, to	direct a flow of ventilation through the workings of the mine by means of air-doors, etc
cowp, to	tip up, overturn
crab	windlass or capstan used in shaft and worked by horses
creep, a	movement within an excavated seam whereby the pressure of the roof overcomes the supports, the floor heaves up, and the floor and the ceiling gradually converge
creeper waste	abandoned area within the coal seam where a creep is taking place, therefore inaccessible and unventilated but liable to contain pockets of gas
cube and fire	brazier with chimney placed over shaft to draw up air from the mine
cuddy	donkey or small horse
dander	rage, panic, loss of control
deputy	underground foreman
drift	underground access passage to meet coal seams or between seams, usually on incline
drive, to	excavate an underground passage

GLOSSARY

dyke	strata slip
engine	colliery pumping or winding steam engine
fathom	depth of six feet
fault	disturbance in the underground strata, usually where a section of a seam, or of seams, has slipped
feeder	underground stream or flow of water
fire, to fire (said of pit)	take fire and explode, to have an explosion of gas
fireman	stoker of colliery engine
fire damp	methane gas (carburetted hydrogen), the source of underground explosions
flags	flagstones, pavement
flay, a	fright
flesh	butcher meat
furnace	fire at foot or head of shaft, to draw air through the workings
galloway	pony or small horse used underground
gin	winch for winding men and materials up and down the shaft, usually horse-powered
gutter	ditch
hay-worm	slow worm
head rig	field furrow
headway	excavated passage within the seam leading to the bords, or faces, opening them up and and providing for ventilation
hewers	face workers, coal-getters
hole, to hole the waste	to break into abandoned workings
hook	attachment for corfs at the end of the winding rope, also forms the loop
horse-keeper	underground stableman

GLOSSARY

in-bye — away from the shaft and towards the face, the opposite of 'out-bye'

incline, inclined plane — sloping portion of waggonway on which the descending full waggons are attached with a rope around a pulley to the ascending empty ones, the latter being thereby pulled up the slope

keel, a — small boat used for conveying coal from the staithe to a ship

kitty — gaol

landsale, a landsale colliery — colliery disposing of coal entirely by land, as distinct from a colliery largely disposing of coal by exporting it on waterways, hence 'seasale'

line, to line (the pit) — survey and measure up the workings

loop — loop formed by hooking back the end of the winding rope, to allow men to ride in the loop

low — candle

lowens — allowance, any bonus or extra payment

marketing — regular domestic purchases at market, particularly of foodstuffs

marrow — workmate

mell — hammer or mallet

main, hence High Main, Low Main — coal seam

monadge — form of box club, from French 'ménage'

necessary, a — privy, private or common latrine, later equivalent being 'netty' (perhaps from 'necessity')

out-bye — at or towards the shaft, the opposite of 'in-bye'

overman — shift foreman, responsible for supervision of underground working

paddock — frog

pawl	device for locking a capstan or winch
penning	perhaps a form of running-way or rail used underground
pillars	blocks of coal left standing between excavated bords to support the roof
plugman	workman in charge of a pumping engine
pore	poker
poke, a	sack or bag
post stone	sandstone
putter	youth pushing or pulling small waggon underground
reasty	rancid
ride, to ride in the shaft	to travel up and down the shaft, attached in some way to the winding rope
rolley	trolley, small wooden truck or waggon for conveying corfs of coal underground, usually only from the face to the waggonway, sometimes the same as 'tram'
rope	winding rope, let down and pulled up the shaft, to transport men and materials
run amain	see 'amain'
scale	draught of air escaping from controlled air-passage, or deliberately allowed through an air-door, but sometimes foul air
scoggan	type of valve-opener, particularly on a Newcomen steam engine
scry, to	spy, notice
seasale	see 'landsale'
shave	woodworker's edge-tool
staithe	short pier on the river, used for tipping coals into vessels
standing, a	apparently a form of partition wall
staple	small shaft sunk underground, sometimes between seams

GLOSSARY

steel mill	device emitting sparks for illumination underground, used in the belief that the sparks would not fire gas
stopping, a	partition wall underground, to direct air flow
sump	drainage pit
swat, to	squat
swill, a	tub
taylor's goose	flat iron
tram, a	see 'rolley', also a waggon for conveying materials other than coal underground, for instance, timber
trouble	break in the coal seam, a 'fault'
viewer	manager of a colliery, both engineering and commercial
waste	worked out and usually abandoned area within an underground seam
wasteman	underground worker who undertakes various maintenance jobs between coal-working shifts, usually not at the face but in the worked-out section of the seam, the waste, and particularly in airways
way	waggonway, or running of a waggonway

INDEX OF PERSONS

This index lists the personal names appearing in the autobiography, notes, introduction and commentaries. When necessary, the spelling of names given in the autobiography is minimally corrected, or else the probable correct spelling is indicated in round brackets. Numbers in bold print refer to the numbered anecdotes and their respective footnotes: other numbers indicate pagination. Page references to named individuals in the footnotes are only included if the individual is not named in the related text. In the case of the Errington family only, references to individuals other than by name (e.g. "his father") are included under the name of the individual (e.g. under Errington, Robert: father of A.E.).

A final reference in square brackets against the name of an individual indicates a source other than the autobiography. An asterisk indicates Heworth chapelry parish register, and CR indicates St Andrew's Catholic chapel register. Note, however, that the Heworth register being unindexed, the references below are only to names noticed while sections of the register were being searched for Erringtons; and also that, while names often appear several times in the register at different dates, the references are normally only to a single exemplary entry.

Anderson, Andrew: Felling collier, killed **43** [*]
Anderson, William: Felling Colliery banksman **28,32,33,149** [* 1799]
Anderson, Dr: Gateshead pill doctor **40,73**,217
Austen, Jane: writer 20
Bailey, George: Felling Colliery plugman **39** [* "collier" 1801 burial, "engineman" 1804 baptism]
Barker, Mr: Newcastle eating house proprietor **41**
Barn(e)s, Thomas: Felling Colliery viewer **11,62**,5,141,149,162
Bell, Andrew: Percy Main collier **57**
Bell, James: Felling Colliery horsekeeper, killed **34** [*]
Bell and Brandling: colliery owners **74**
Belly, Matthew: Felling schoolboy **4** [cf. * "Belly" 1798 baptism]
Bewick, Thomas: engraver 1,22,260,261
Bilton, - : prize-fighter **41**
Bilton, Tom: schoolboy, son of William **26**,184
Bilton, William: Newcastle millwright **26**
Blenkinsop, John: engineer 4,5,14,142
Bowman, Mrs: Benwell widow and landlady **48**,170
Bowman, Anthony: Morpeth farmer, son of Mrs Bowman **48**
Brandling, Mr: colliery owner **43,50,74,76**,163,165
Brandling, Charles: colliery owner 2
Brandling, John: colliery owner **49**
Brandling, William: colliery owner [R.W. Brandling] **74**,166,
Brandlings: colliery owning family 4,13,14,55,142,162,188,191
Brown, John: Fawdon shave-maker **76**
Brown, Ralph: Felling Colliery wasteman **32**

INDEX OF PERSONS

INDEX OF PERSONS

INDEX OF PERSONS

Marshal, John: Percy Main Colliery wasteman **62**
Marshal, Richard: Coxlodge enginewright **76**
Miller, Robert: Felling innkeeper **37**
Miller, the: prize-fighter **41**
Morchents, Mary: Newcastle barmaid **29**
More, Hannah: author 26,183
Morley, Christopher: Felling Colliery workman **11** ["engineman" * 1800 baptism]
Morley, Sopwith: Felling Colliery engineer **11** ["engineman" * 1795 baptism]
Nisbet, Thomas: Heworth schoolboy **4**
Oliver, John: Percy Main Colliery overman **61**
Oliver, Mr: Backworth Colliery viewer **78,81**,165
Oxberry, John: local historian 230,232
Percys: dukes of Northumberland 14,84
Purves, Charles: Felling Colliery clerk **31**
Ray, William: Felling freemason and atheist **44** [? "William Reay" * 1795 baptism]
Richardson, Jane: mother of a child by A.E. 99,109,124,170,178
Riddly (Ridley), George: Felling Colliery hewer, killed **43**,156,191 [*]
Robison, Thomas: Gateshead Fell tailor **17,25**
Robson, Jane: Felling housewife **36**
Robson, Mr: Felling Colliery agent **43,55**,165
Rodgers (Rogers), Edward: Felling Colliery overman **43** [* 1800 baptism]
Rodgers (Rogers), William: Felling Colliery furnaceman **14,19**
Roe, Thomas: Heworth resident **7**
Russell, Squire: of Low Heworth **22**
Sanderson, Matthew: Felling collier **29** [killed 1812, aged 35, Hodgson 1813]
Sill, Mr: steward to the Ellisons of Gateshead Park House **31**,175 ["Thomas Sill, Kirton's Gate, Gateshead, Land Steward", aged 87 * 1801 burial]
Smiles, Samuel: writer 9
Smith, Joseph: Coxlodge workman (?) **77**
Solsby, Padison: waggonway wright (?) **47**
Stephenson, George: railway engineer 2,4,5,14,22,143,146
Stevenson, Mark: Felling Colliery fireman **39**
Stewart (Stuart), John: Fawdon collier, Roman Catholic **70,71**,195 [CR]
Stobbs, Henry: farmer at Carr Hill **21** [* 1794 baptism]
Stove, Robert: Felling Colliery deputy **32**,160 [rescued 1812, Hodgson 1813, killed 1813 *]
Straker, John: Felling Colliery viewer **32,39,43,49,50,55,83**,149, 162,163,165,174 [* 1802 baptism]
Taylor, John: Percy Main Colliery banksman (?) **55**
Taylor, Mr: Thomas Taylor of Backworth, colliery owner **80,81,83**, 15,164,166
Thobren (Thowburn), Mrs: school mistress **4,5,9** [* 1787 burial]
Thompson, John: Felling Methodist 19

Trumble, James: Felling collier **43**,195 [in rescue 1812, Hodgson 1813]
Trumble, William: Felling Colliery horsekeeper 24 [* 1801 baptism, CR]
Turnbull, brothers: Newcastle shopkeepers **51**
Urwin, Nickell: Felling Colliery deputy **32,33**,160 [killed 1812, aged 58, Hodgson 1813]
Wake, Mr: of Southwick **46**
Warlow (Warrilow), Revd. William: Roman Catholic priest **31**,196
Welford, Richard: local historian 230,232,233
Wesley, John: preacher 10
Wilson, Matthew: Felling Colliery plugman **39**
Wood, Joseph: edge tool maker **47**
Worker, George: farmer, near Hebburn **53**
Yellowly, William: Heworth schoolmaster **5,7,8,10** [*]

INDEX OF PLACES

This index lists the place-names appearing in the autobiography, the notes, the introduction and the commentaries. Streets, buildings (including churches), pits and other minor localities within settlements are not listed. Numbers in bold print refer to the numbered anecdotes and their respective footnotes; other numbers indicate pagination. Page references to places mentioned in footnotes are only included when the place is not referred to in the related text and when the information is substantial.

INDEX OF PLACES

INDEX OF ANECDOTES

This index lists the numbers of those pages in the Commentaries that contain a reference or references to each numbered anecdote.

INDEX OF ANECDOTES

ILLUSTRATIONS IN THE TEXT

The background illustrations in the text are mainly a selection from the renowned tailpieces by Thomas Bewick, since these happen to illustrate the lower Tyne and many aspects of the lives of its people in A.E.'s lifetime - albeit aspects above ground only (despite Bewick's father owning a small coalmine). Engraved by Bewick and his pupils between the 1780s and 1810s, in order to variegate the interest of the natural history books for which Bewick was providing copious illustrations, the tailpieces were assembled in his **Vignettes** (Newcastle upon Tyne, 1827), with a few more appearing in the 1885 memorial edition of his works.

Some of the details of those reproduced here deserve comment. The smoking chimneys frequently shown in the background must be, more often than not, those of colliery steam-engines; and the distinctive spire of St Nicholas' church in Newcastle also sometimes appears - for instance, in the background to a scene of children playing in a stream (p.117). The high skyline topped with windmills seen across a river from a stone inscribed with the coat of arms of Newcastle (p.138) presumably represents the Windmill Hills at Gateshead, while the steep cliffs by the seaside (p.138) are at Tynemouth. Ships being loaded with coal at a staithe (p.21) are in 'Coaly Tyne'. According to Bewick, the ruined chapel (p.117) was the Roman Catholic chapel in High Street, Gateshead, fired by a mob in 1745. The funeral depicted (p.125) was at Ovingham church, where Bewick was himself later to be buried. The demobilised soldier (p.181) recalls A.E.'s brother, George; and the gibbet (p.220), the dog (p.ii), the cuddy (p.i), the suicide (p.229), and the night-fright (p.237) represent those that appear in A.E.'s anecdotes. A drunk staggers home (p.214); while rude health is invoked in a view of a 'netty' or privy in use (p.217). A child plays with a home-made toy waggon (p.254), as A.E. did; a boy goes bird-nesting (p.220), as A.E. did too, but not as disastrously. A man directs five horses along a road, perhaps after a horse fair (p.248); a cart with a single horse has to stand in as the nearest thing to A.E.'s 'Manchester waggon and eight horses' (p.259). Twice A.E. rescued a traveller in the snow (p.261).

Bewick's vignettes are usually celebrated for their humour and rustic charm, but his own comments on them illuminate their often enigmatic character (see Appendix 1 of the 1978 reprint of the **Vignettes**, helpfully edited by Iain Bain). Most of the tailpieces were in fact intended to point a moral; and the moralising is not uncommonly dismissive of human frailty, jaundiced, embittered and even cynical - a stage further away from A.E.'s frame of mind than the genial enthusiasm for reformist ideas expressed in the 'Memoir' Bewick penned in old age. Perhaps the tailpieces with their black humour enabled the young engraver to work off the constraints and frustrations inherent in the production of factual representations of birds and quadrupeds. However, although the concealed moral is, at times, inappropriate to A.E.'s autobiography, the superficial details of the selected vignettes, particularly those details to which attention is drawn in this note, are not. Hence the list below describes each tailpiece in terms of its relevance to the autobiography and does not always indicate Bewick's intention or repeat his explanation.

LIST OF ILLUSTRATIONS IN THE TEXT (B = Bewick)

The point of tailpieces is that they catch the eye by appearing below a chunk of print, singly. The use of Bewick's tailpieces as illustrations in this work has meant that some have lost artistic validity and value, not only by the obvious difficulty of reproduction, but by being doubled on a page. In apology to readers thereby offended, I conclude with a single and particularly fine tailpiece.

ACKNOWLEDGEMENTS

I came upon A.E.'s autobiography and transcribed it in 1950, while a research student at Nuffield College, Oxford. Thereafter my notes accompanied me to distant lands, until they settled into a Merseyside drawer. The text of the autobiography was typed first and last by myself, but in between at least four times, in different versions of editing, by typists at Berwick-on-Tweed, Ibadan, Freetown and Liverpool - whose labours in respecting A.E.'s spelling I now salute. In my youth I was urged to proceed with an edition by G.D.H. Cole and Kenneth Connell; in senescence I have been prodded by Norman McCord, Keith Wrightson and David Vincent.

I would have been unable to produce the camera-ready copy for this edition without the guidance of my experienced colleagues, Michael de Cossart and Patrick Tuck, or without the advice of Robin Bloxsidge of the Liverpool University Press; and I thank the Department of History for its publication of the work. Another colleague, Philip Bell, gave up a slice of his time in order to read and proofread the edition, and it owes much to his meticulous care. Unfailing moral support, not least in the form of regular cups of coffee laid beside the WP, has come from my wife, Margaret Hair.

For aid in the search for additional documentation I am indebted to archivists and librarians at Newcastle Central Library and Durham and Northumberland County Record Offices, as well as to those at national libraries and archives. Information on Gateshead and Felling in the Local History Collection of Gateshead Central Library was generously brought to my notice and put at my disposal by Mr Michael Henry, Mr Tom Marshall and Miss Edith Carnaffin; and information on Tynemouth and North Shields I owe largely to Mr E.J.Hallerton of the Local Studies Centre of North Shields Central Library, who excelled in prompt and full replies to postal queries. I thank the Library Committee of Gateshead Metropolitan Borough Council for allowing me to proceed with a published edition of the Errington manuscript, and Mr Patrick Conway, the Borough Librarian, for his assistance in gaining me this permission. Many years ago, Father Edward Harriott, then parish priest of St Andrew's Catholic church, Newcastle, consulted the church's unpublished registers on my behalf; recently Mr John McKie drew on NCB records to inform me about earlier Felling pits; and Mrs Jennifer Duncan has checked Felling census material in London. My daughter, Ruth Hair, looked up records in Newcastle and nobly shared her student room with me when I visited the North East. I thank them all. Finally, I record with deep appreciation that most of the information about Felling topography has come from Mr Maurice Armstrong, a retired railway engineer, whose knowledge of the history and records of industrial Felling has enabled me to establish many points in more minute detail than I had ever thought possible before I met him. Mr Armstrong undertook hours of library research into the Felling sites mentioned by A.E., and on 14 December 1987 drove me around them and showed me the lie of the land. I only hope that the use I have made of the information supplied by Mr Armstrong, and bv all the other individuals mentioned above, is worthy of their efforts. If I have misunderstood or distorted their information, the fault is mine, and I bear responsibility for all errors. P.E.H.H.

ILLUSTRATIONS

DIAGRAMS

The diagrams and explanations of coalmining technology that appear on the five following pages are extracted from John Holland, **The history and description of fossil fuel, the collieries and the coal trade of Great Britain,** London, 1835 (2nd. edition 1841, reprinted 1968). Although not always correct in every detail, they present a simple and clear picture of certain important features of Tyneside coalmining in the early decades of the nineteenth century. The items described are:

On p.270, the plan of Felling Colliery in 1812 is from Hodgson 1813.

OTHER ILLUSTRATIONS

The illustrations taken from T.H.Hair, **Sketches of the Coal Mines in Northumberland and Durham** (author's title-page 1839, published London 1844) were drawn in the mid 1830s, in the final years of A.E.'s working career. Having appeared originally in a folio volume, their reduction in size for reproduction here has considerably reduced their quality. Their original titles were as follows: 'The A pit, Fawdon Colliery', 'Percy pit, Percy Main Colliery', 'Bottom of the shaft, Walbottle Colliery', 'Crane for loading the rollies'.

 are called *troubles,* or *faults,* from their troubling, or putting to fault, the pitmen; or the latter term may have arisen from the supposition that the rents have been occa- [*A*]
sioned from something faulty in the aggregation of the matter of the rocks themselves. The subjoined diagram, which will strikingly illustrate the disloca-

Fig. 18.

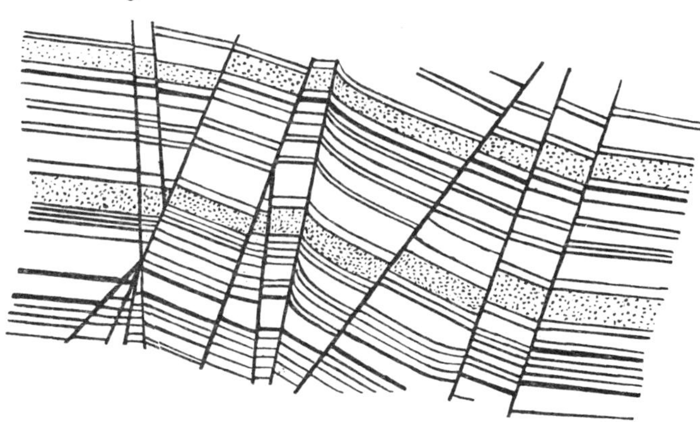

tions in question, is from a splendid section, on a large scale, presented by Mr. Buddle to the Natural History Society of Newcastle-upon-Tyne, in which neighbourhood (at Jarrow) the portion of coal measures thus singularly broken up, occurs.

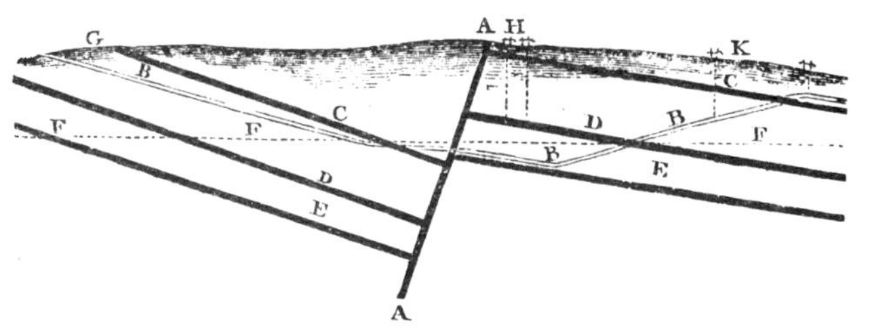

[*B*]

The above diagram exhibits a section of the great slip as it cuts a portion of the coal field near Newcastle:

A A. The Ninety-fathom Dyke, of unknown depth.

C C. The High Main coal.—D D. The Low Main coal.—E E. The Beaumont seam.—F F F. Level of the river Tyne.—

[*C*]

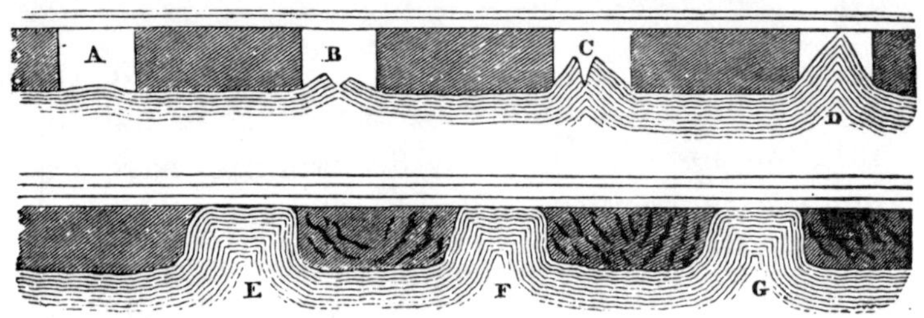

A creep is a certain protrusion or bursting upwards of the floor of a mine, owing to the small extent, and consequent liability to sink, of the bases of those pillars of coal which are left to support the roof during the excavation of so much of the seam as appears compatible with safety to the pitman.

The annexed cut represents the exterior appearance of one of the old-fashioned steam-engines, still very common about collieries, and called a whim,

Fig. 25.

[*D*]

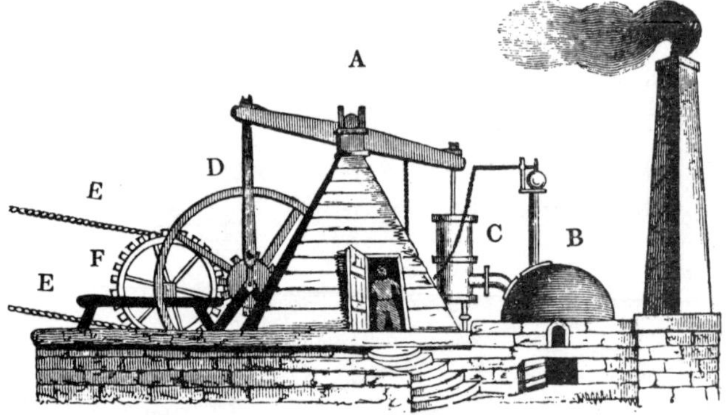

or whimsey (*fig.* 25). A the engine-beam. B the boiler. C the working cylinder. D the fly-wheel. F the cog-wheel, carrying the drum upon which the rope is wound. E E the ropes, passing to the head-gear (*see fig.* 23). In the working of this machinery, the engine-tenter stands with his hand upon a lever, to stop the steam, and stay the action of the whole, the moment he sees the corfe aboveground.

The annexed slight profile sketch of the works at Carville on the Tyne, in the Wallsend group of collieries, will give some idea of the arrangements above

Fig. 26. [*E*]

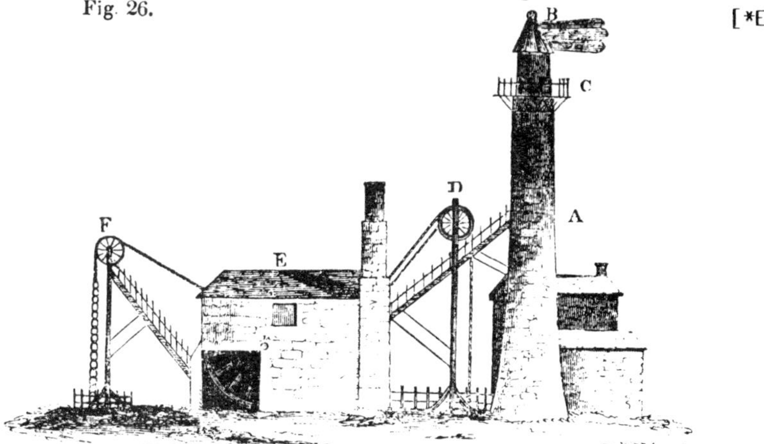

described (*fig*, 26). A the tall brick funnel adjacent to the upcast shaft. B the smoke disperser. C a platform for convenience of cleansing, repairs, &c. D head-gear (*fig*. 24). over the drawing pit, supporting the wheels over which the ropes pass. E engine-house, containing the fly wheel, winding cylinders, and other machinery. F counterpoise.

[*F*]

It is not uncommon to see a horse yoked *behind* the loaded waggons in their descent; the animal by this means assisting to prevent their too rapid motion, and being also ready to draw the waggons back, after they have discharged their contents at the staithes.

In practice, the following is the method of ventilation ordinarily adopted;—A (*fig.* 29.) is the down-

Fig. 29.

[*G*]

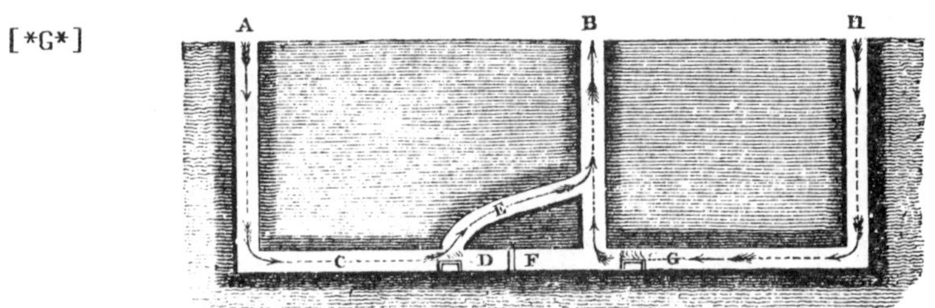

cast shaft, and B the upcast; at a short distance from the bottom of the latter, and in the connecting passage C, is placed a furnace D, consisting of a platform of iron bars, raised somewhat from the ground, and covered with a fire of 7 or 8 feet in width, by 12 feet in length. The smoke and draught of this fire, instead of being connected with the bottom of the pit B, are provided for by the carrying of an arched drift E, in an inclined direction, from above the fire-place into the shaft, at a little distance from the bottom.* A stopping of boards is placed at F, to prevent access of the air in that direction. In what is called compound ventilation, a passage is carried from G, where there is also a furnace, to another downcast pit H, in an opposite direction to the first. Wallsend, Percy Main, Hebburn, and Heaton collieries, were all ventilated upon this principle.

[*H*]

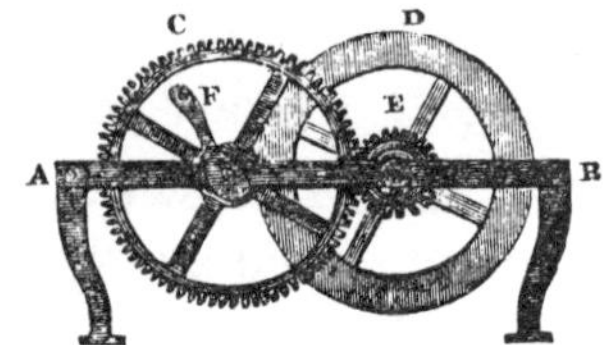

The figure annexed is a side view of a steel mill of simple construction.

Suppose the annexed diagram (*fig.* 32.) to represent a field of coal: the oblong dark shaded portions will shew the pillars, and the white spaces the extent from which the coal was dug out, in the manner described in the preceding Chapter, and before the working of the pillars was attempted. The whole of these pillars were considered necessary for securing the roof, and no attempt was made to reduce them, previously to 1795, from the apprehension of producing a creep. In that year, when the Walker Colliery became exhausted, as to the main seam, an attempt was made at partial working, by removing one half of every alternate pillar, as shewn in the cut. By this means 55, instead of only 40, per cent. of the entire coal was obtained; the remaining 45 per cent. being still left in the pillars, and consequently lost. The double lines in the transverse openings represent the air stoppings of brick-work, and the darts shew the direction of the current of air.

Fig. 32.

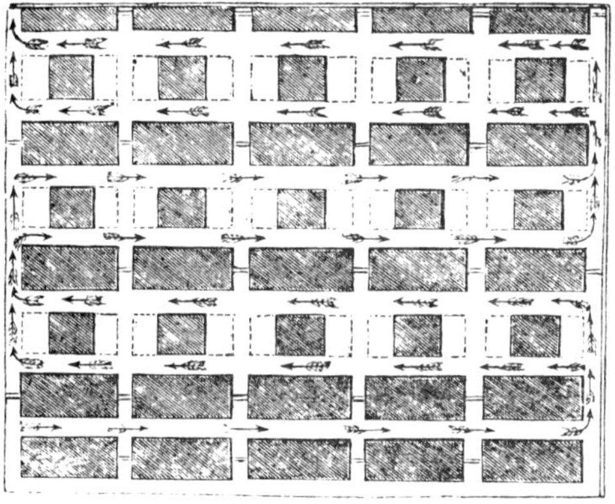

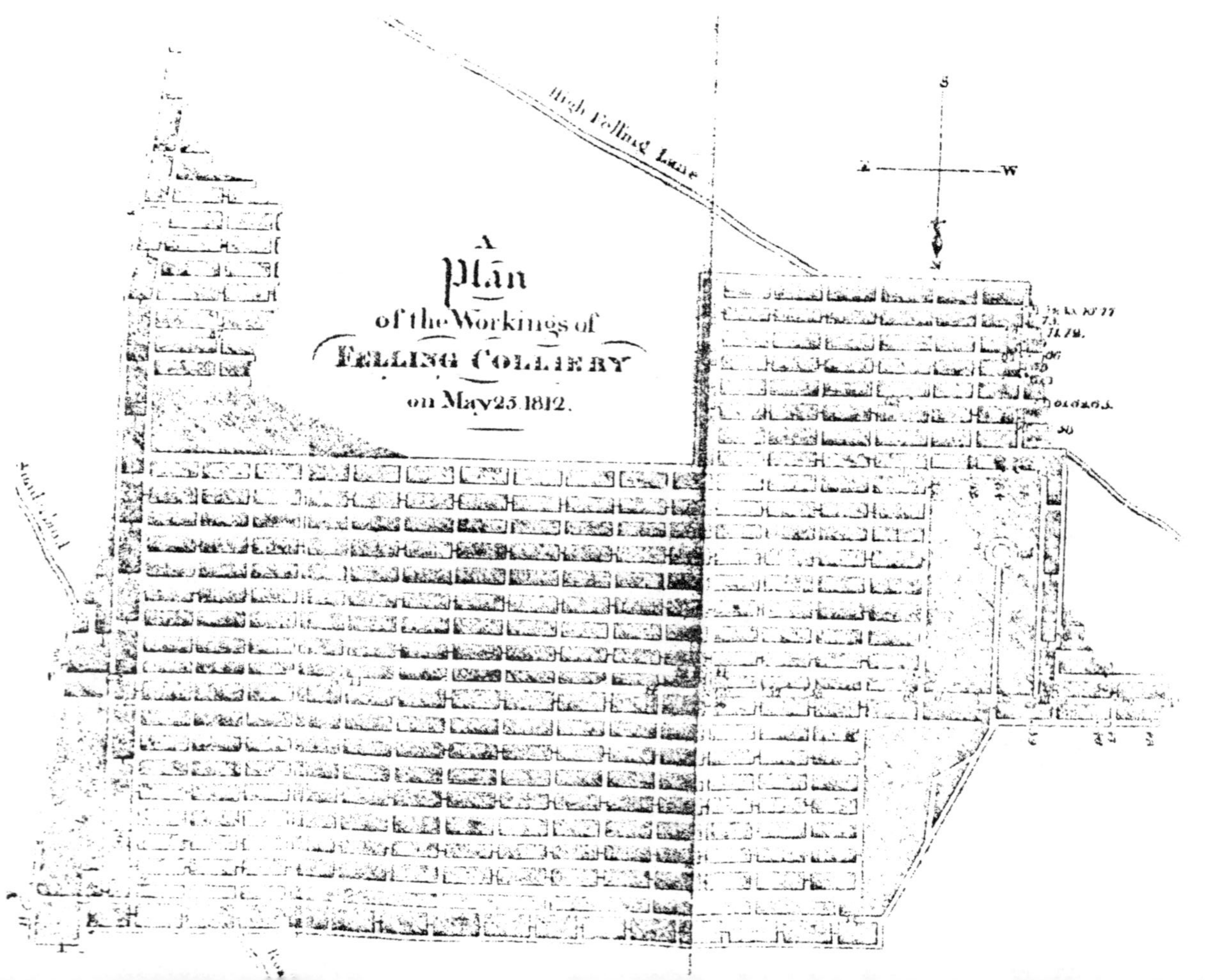
A Plan of the Workings of FELLING COLLIERY on May 25 1812.
High Felling Lane
S
W

TYNESIDE SURFACE WAGGONWAYS c.1770

FAWDON COLLIERY (A Pit) c.1835

PERCY MAIN COLLIERY (Percy Pit) c.1835

A SHAFT BOTTOM

AN UNDERGROUND CRANE

(above) GATESHEAD SEEN ACROSS THE TYNE BRIDGE FROM NEWCASTLE c.1835

(below) CHEMICAL WORKS AT FRIARS GOOSE c.1850 (Felling left background)

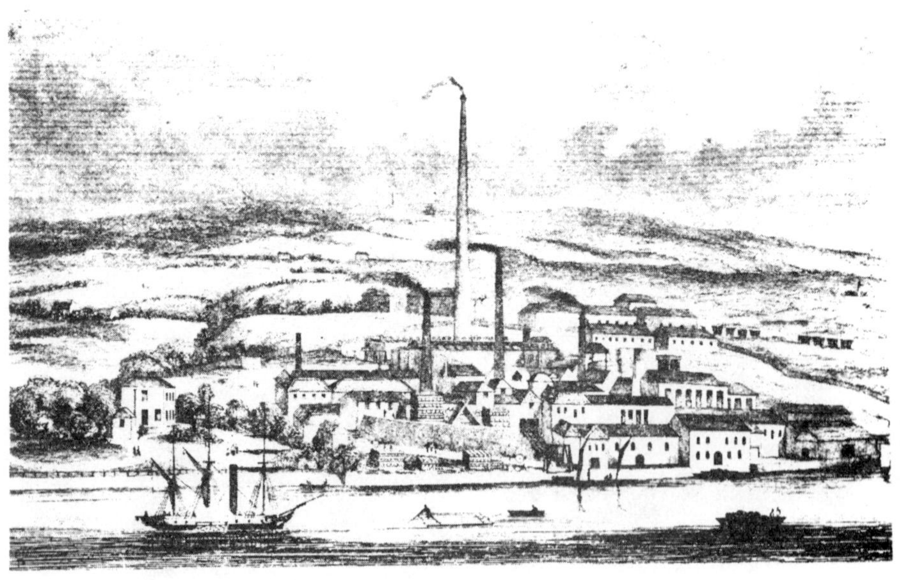

OLD GATESHEAD - houses in HIGH STREET c.1890

Register of the Births of the fol:
lowing Sons of [illegible] Errington within
the Chapelry of Heworth

Wm Son of Robert Errington Augt 10th 177[illegible]
John Son of Robert Errington Sept 24th 177[illegible]
George Son of Robert Errington Octr 6th 177[illegible]
Anthony Son of Robert Errington Sept 20th 1778

Heworth, Robert born Octr 29th 1 Son of Anthony
Errington of this Chapelry by his wife
Ann late Hindmarsh native of Kirkwhelping[illegible]

ENTRY OF ERRINGTON BIRTHS IN HEWORTH CHAPELRY REGISTER
(by permission of the Vicar of Heworth and Durham County Record Office)

BODIES BEING BROUGHT TO THE SURFACE AND COFFINS BEING UNLOADED
(enlarged illustration from the title-page of
'Narrative of a Dreadful Occurrence at Felling Colliery, On the 25th of May, 1812.'
(Religious Tract Society, Liverpool)

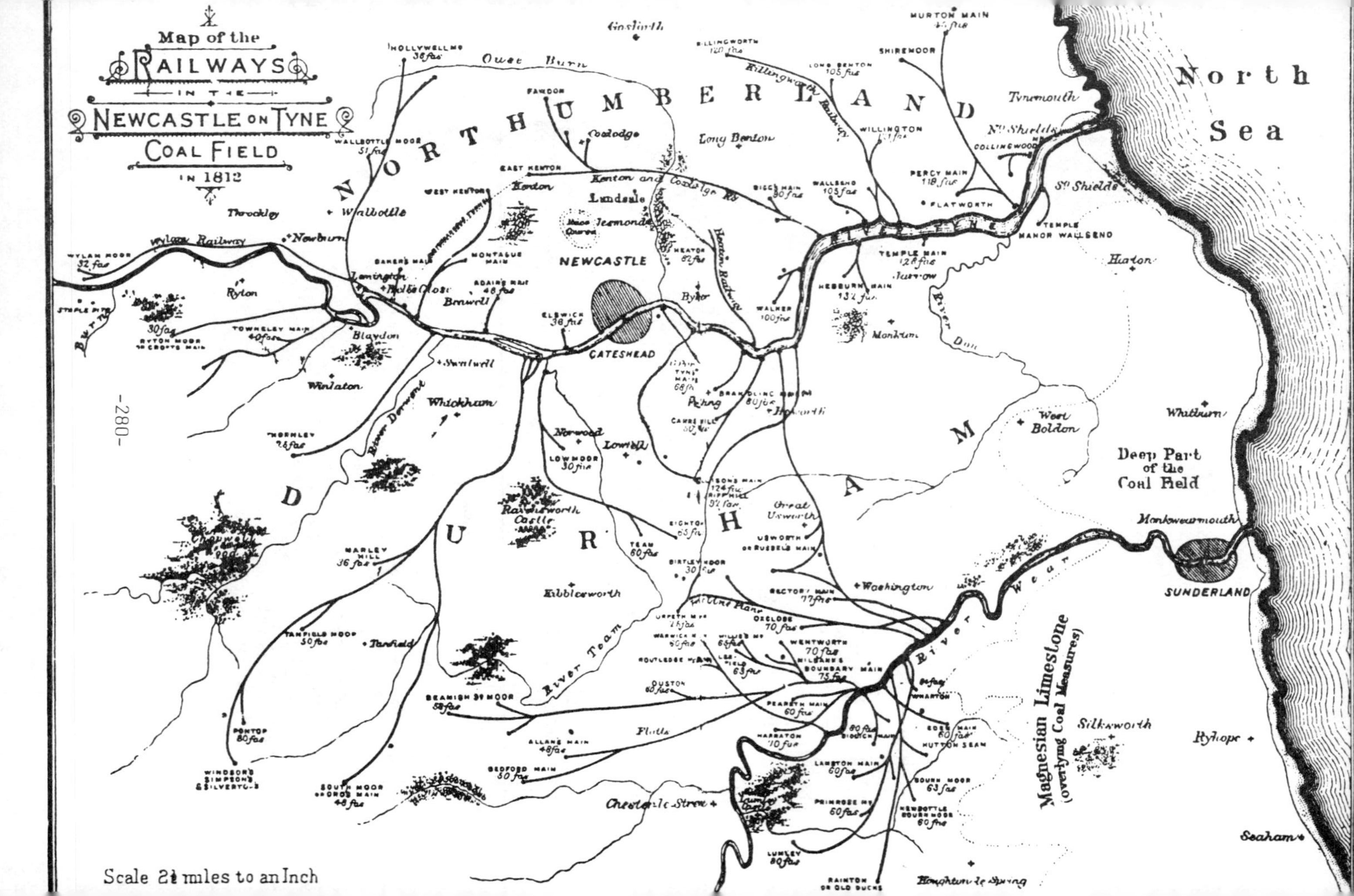
Map of the
RAILWAYS
IN THE
NEWCASTLE ON TYNE
COAL FIELD
IN 1812
Scale 2½ miles to an Inch
North Sea
NORTHUMBERLAND
DURHAM
NEWCASTLE
GATESHEAD
SUNDERLAND
Deep Part of the Coal Field
Magnesian Limestone (overlying Coal Measures)
Kenton and Coxlodge Rs
Heaton Railway
Killingworth Railway
Wylam Railway
River Derwent
River Team
River Wear
River Don
Ouse Burn
Tynemouth
Nth Shields
Sth Shields
Monkwearmouth
Seaham
Ryhope
Silksworth
Whitburn
West Boldon
Washington
Chester le Street
Houghton le Spring
Gosforth
Long Benton
Jesmond
Whickham
Blaydon
Winlaton
Ryton
Newburn
Throckley
Walbottle
Kibblesworth
Ravensworth Castle
Lumley Castle
WYLAM MOOR 32 fas
HOLLYWELL MN 38 fas
WALLBOTTLE MOOR 51 fas
FAWDON
EAST KENTON
WEST KENTON
MONTAGUE MAIN
ADAIRS MAIN 48 fas
ELSWICK 38 fas
HEATON 82 fas
BIGGS MAIN 80 fas
WALLSEND 105 fas
WILLINGTON
PERCY MAIN 118 fas
FLATWORTH
MURTON MAIN
SHIREMOOR
KILLINGWORTH 120 fas
LONG BENTON 105 fas
COLLINGWOOD
TEMPLE MAIN 128 fas
TEMPLE MANOR WALLSEND
HEBBURN MAIN 132 fas
WALKER 100 fas
LOW MOOR 30 fas
TEAM 60 fas
BIRTLEY MOOR 30 fas
USWORTH OR RUSSELS MAIN
RECTORY MAIN 77 fas
OXCLOSE 70 fas
WENTWORTH 70 fas
BOUNDARY MAIN 75 fas
PEARETH MAIN 60 fas
HARRATON 110 fas
LAMBTON MAIN 60 fas
PRIMROSE HY 60 fas
LUMLEY 80 fas
RAINTON OR OLD DUCKS
BOURN MOOR 63 fas
NEWBOTTLE BOURN MOOR 60 fas
HUTTON SEAM
WHARTON
OUSTON 60 fas
BEAMISH 3D MOOR 58 fas
ALLANS MAIN 48 fas
BEDFORD MAIN 80 fas
SOUTH MOOR GEORGE MAIN 48 fas
WINDSORS SIMPSONS & SILVERTOPS
PONTOP 80 fas
TANFIELD MOOR 50 fas
Tanfield
MARLEY HILL 36 fas
WORMLEY 28 fas
TOWNELEY MAIN 40 fas
RYTON MOOR 30 fas
STAPLE PITS

POSTSCRIPT ADDITIONS AND CORRECTION

p.135 As this edition was going to the press, Mr Maurice Armstrong was shown a copy (in private possession) of a booklet printed in 1927 containing a brief but useful history of the Catholic churches and schools at Felling (**Souvenir Handbook of the Foundation Ceremony of** St. Alban's **Catholic School, Pelaw,** Amos Almond, Printer, Felling, 28 pp.). The history includes the statement that, in the late 1850s, St John's School was built in "a field in High Street, reputed to be part of the Brandling Estate; for it was at one time attached to the farm of Robert Errington, woodman to the Brandlings". Later, St Patrick's Catholic Church was built "on the higher level" of the same field (p.6). Since the former school building and the present church stand in the lower part of High Street, Felling, the Errington field lay in what was earlier termed Low Felling; and this would confirm my supposition about the location of the farm. The anonymous author of the 1927 booklet thanked Mr John Oxberry for providing him with "invaluable local data" (p.16), and since the Errington manuscript was then in the possession of Oxberry, no doubt it was he who contributed the reference to Robert Errington and his farm. However, A.E.'s autobiography does not in fact locate the farm, and another source than the autobiography must have supplied the information that the school and church were built on a field "at one time" belonging to the farm. Mr Armstrong has traced in the Catholic diocesan archives at Newcastle an indenture of 1909 referring to a deed of 1867 by which the church authorities obtained additional land at Felling, presumably for the church. Part of the land was described as "formerly parcel of the Fell moor or common ... commonly called ... Holly Hill formerly in the possession tenure or occupation of Robert Errington and John Pattison as tenants thereof". This, or an earlier legal document concerning the same land, was almost certainly Oxberry's source. The author of the 1927 booklet thanked a number of person, apparently parishioners, for "contemporary records and a wealth of detail", and among these was a Mr Geo. Errington. It is therefore possible that descendants of A.E. lived at Felling in the 1920s, knew something of the family history, and pointed Oxberry in the direction of the documentation of the church site. Oxberry was a regular contributor of notes on local history to the Tyneside press, and a note by him on the Erringtons may yet be traced, although his collection of press cuttings at Gateshead Central Library does not appear to contain one.

p.161, "no-one ever pointed the finger at Buddle". Correction: the eccentric William Martin certainly did (Balston [p.22,n.2], pp.133-4).

p.230 Add to the reference to David Vincent's book that he has listed the autobiography in J.Burnett, D.Vincent, and D.Mayall, **The autobiography of the working class, I** (Brighton, 1984), p.104.